Lectures in Mathematics
ETH Zürich
Department of Mathematics
Research Institute of Mathematics

Managing Editor:
Michael Struwe

Vladimir Turaev
Introduction to Combinatorial Torsions

Notes taken by Felix Schlenk

Springer Basel AG

Author's address:

Institut de Recherche Mathématique Avancée
Université Louis Pasteur – CNRS
7 rue René Descartes
67084 Strasbourg Cedex
France

2000 Mathematical Subject Classification 57Q10, 57R57, 19J10, 57M25

Library of Congress Cataloging-in-Publication Data

Turaev, V.G. (Vladimir G.), 1954–
  Introduction to combinatorial torsions / Vladimir Turaev ; notes taken by Felix Schlenk
      p.   cm. -- (Lectures in mathematics ETH Zürich)
  Includes bibliographical references.

  1. Manifolds (Mathematics).   2. Torsion theory (Algebra).   3. Homology theory.   I. Title.
  II. Series.
  QA613.T87   2000
  516'.73 – dc21

Deutsche Bibliothek Cataloging-in-Publication Data
Turaev, Vladimir G.:
Introduction to combinatorial torsions / Vladimir Turaev. Notes taken by Felix Schlenk. -
Basel ; Boston ; Berlin : Birkhäuser, 2001
  (Lectures in mathematics)
  ISBN 978-3-7643-6403-8   ISBN 978-3-0348-8321-4 (eBook)
  DOI 10.1007/978-3-0348-8321-4

© 2001 Springer Basel AG
Originally published by Birkhäuser Verlag in 2001

Member of the BertelsmannSpringer Publishing Group
Printed on acid-free paper produced from chlorine-free pulp. TCF ∞

9 8 7 6 5 4 3 2 1

Contents

Introduction

This book is an extended version of the notes of my lecture course given at ETH in spring 1999. The course was intended as an introduction to combinatorial torsions and their relations to the famous Seiberg–Witten invariants.

Torsions were introduced originally in the 3-dimensional setting by K. Reidemeister (1935) who used them to give a homeomorphism classification of 3-dimensional lens spaces. The Reidemeister torsions are defined using simple linear algebra and standard notions of combinatorial topology: triangulations (or, more generally, CW-decompositions), coverings, cellular chain complexes, etc. The Reidemeister torsions were generalized to arbitrary dimensions by W. Franz (1935) and later studied by many authors. In 1962, J. Milnor observed that the classical Alexander polynomial of a link in the 3-sphere S^3 can be interpreted as a torsion of the link exterior. Milnor's arguments work for an arbitrary compact 3-manifold M whose boundary is non-void and consists of tori: The Alexander polynomial of M and the Milnor torsion of M essentially coincide.

My own interest in torsions was inspired by a question of O. Viro (1972) concerning mutual relations between the Alexander polynomials of links in a compact 3-manifold. In 1975, building on the paper of Milnor, I obtained an equivalence between the Alexander polynomial and the Milnor torsion for closed 3-manifolds. In 1976, I introduced a more powerful torsion called in this book the "maximal abelian torsion". It contains the Milnor torsion as an ingredient but in general bears more information. The study of the maximal abelian torsion constituted the bulk of my Ph. D. thesis (1979).

The Reidemeister torsions of a manifold M have a well-known indeterminacy. They are defined only up to multiplication by ± 1 and elements of $H_1(M)$. The study of this indeterminacy led me in the 1980's to two new combinatorial structures on manifolds (of any dimension): homology orientations and Euler structures. In the presence of these additional structures, the indeterminacy in the definition of torsions can be completely eliminated. This gives rise to so-called sign-refined torsions and to the torsions of Euler structures. In particular, the sign-refined version of the Milnor torsion computes the Conway link polynomial.

In 1996, Meng and Taubes discovered a remarkable connection between the Alexander polynomial of a closed oriented 3-manifold M and the Seiberg–Witten invariant SW_M of M. They proved that the Alexander polynomial of M can be explicitly computed from SW_M. Their proof uses the theory of sign-refined torsions mentioned above. Using the results of Meng and Taubes and capitalizing on the theory of Euler structures, I showed that the Seiberg–Witten invariant SW_M is equivalent (at least up to sign) to the refined maximal abelian torsion of M. This yields a combinatorial computation of the

Seiberg–Witten invariant in dimension 3 (up to sign). One of the key points is that the Euler structures on a 3-manifold M canonically correspond to the Spin^c-structures on M appearing in Seiberg–Witten theory. Note that a combinatorial computation of the Seiberg–Witten invariant in dimension 4 is an outstanding open problem.

The book consists of three chapters. Chapter I presents the algebraic foundations of the theory of torsions. Chapter II is concerned with various versions of the Reidemeister torsion of CW-complexes and manifolds. The Milnor torsion and the maximal abelian torsion are discussed in detail as well as their relations with the Alexander polynomial. This chapter includes a certain amount of standard material on the basics of algebraic topology, simple homotopy theory, torsions of lens spaces, knots and links, Fox differential calculus, etc. This material was included in my lectures to make it more accessible to beginners. I decided to keep it in the book for the same reason. Chapter III is concerned with homology orientations, Euler structures, and refined torsions. At the end of Chapter III, the connections between torsions and the Seiberg–Witten invariant of 3-manifolds are briefly described.

I end the introduction with two remarks. First of all, this book is not a systematic treatise on torsions or on simple homotopy theory. For this, the reader is referred to Milnor's excellent survey [24], to Cohen's monograph [6], and to my own survey [34]. The analytical aspects of the theory of torsions are completely left out, see [27, 4]. Secondly, the connections between torsions and the Seiberg–Witten invariant can be exploited to study both invariants. I hope to discuss this in more detail elsewhere.

The notes of my lectures, which served as the basis for this book, were taken by Felix Schlenk. His collaboration was crucial at all steps of the preparation of the book. Among other things, he drew all the figures. I would like to use this opportunity to thank Felix for his enthusiasm and hard work.

Chapter I
Algebraic Theory of Torsions

1 Torsion of chain complexes

Let $\mathbb{F}$ be a field and let D be a finite-dimensional vector space over $\mathbb{F}$. Suppose that $\dim D = k$ and pick two (ordered) bases $b = (b_1, \ldots, b_k)$ and $c = (c_1, \ldots, c_k)$ of D. Then

$$b_i = \sum_{j=1}^{k} a_{ij} c_j, \qquad i = 1, \ldots, k,$$

where the transition matrix $(a_{ij})_{i,j=1,\ldots,k}$ is a non-degenerate $(k \times k)$-matrix over $\mathbb{F}$. We write

$$[b/c] = \det(a_{ij}) \in \mathbb{F}^* = \mathbb{F} \setminus \{0\}.$$

Clearly,

1) $[b/b] = 1$,
2) if d is a third basis of D, then $[b/d] = [b/c] \cdot [c/d]$.

We call two bases b and c *equivalent* ($b \sim c$) if $[b/c] = 1$. The properties 1) and 2) show that $\sim$ is indeed an equivalence relation.

Let

$$0 \to C \hookrightarrow D \xrightarrow{\beta} E \to 0$$

be a short exact sequence of vector spaces. Then $\dim D = \dim C + \dim E$. Let $c = (c_1, \ldots, c_k)$ be a basis of C and $e = (e_1, \ldots, e_l)$ be a basis of E. Since β is surjective, we may lift each e_i to some $\tilde{e}_i \in D$. We set

$$ce = (c_1, \ldots, c_k, \tilde{e}_1, \tilde{e}_2, \ldots, \tilde{e}_l).$$

Then ce is a basis of D. Its equivalence class does not depend on the choice of $\tilde{e}_i$. It depends only on the equivalence classes of c and e.

Let $C_0, C_1, \ldots, C_m$ be finite-dimensional vector spaces over $\mathbb{F}$ and

$$\partial_i \colon C_{i+1} \to C_i$$

be linear homomorphisms. The sequence of vector spaces and homomorphisms

$$C = (0 \to C_m \xrightarrow{\partial_{m-1}} C_{m-1} \to \ldots \xrightarrow{\partial_1} C_1 \xrightarrow{\partial_0} C_0 \to 0)$$

is called a *chain complex* of length m if $\operatorname{Im} \partial_i \subset \operatorname{Ker} \partial_{i-1}$ for all $i = 1, \ldots, m$. This condition is equivalent to $\partial_{i-1} \circ \partial_i = 0$ for all i. The vector space $H_i(C) = \operatorname{Ker} \partial_{i-1} / \operatorname{Im} \partial_i$ is called the *i-th homology* of the chain complex C.

Definition 1.1 The chain complex C is said to be *acyclic* if $H_i(C) = 0$ for all i. (This holds if and only if $\operatorname{Ker} \partial_{i-1} = \operatorname{Im} \partial_i$ for all i.) The chain complex C is said to be *based* if each C_i has a distinguished basis c_i.

While in homology theory one studies homology of spaces and chain complexes, the theory of torsions deals mainly with acyclic chain complexes. The subtle point is to produce acyclic chain complexes in a topologically relevant way.

Let $C = (0 \to C_m \to \ldots \to C_0 \to 0)$ be an acyclic based chain complex over $\mathbb{F}$. Set $B_i = \operatorname{Im}(\partial_i \colon C_{i+1} \to C_i) \subset C_i$. Since C is acyclic,

$$C_i / B_i = C_i / \operatorname{Ker}(\partial_{i-1} \colon C_i \to C_{i-1}) \cong \operatorname{Im} \partial_{i-1} = B_{i-1}.$$

In other words, the sequence

$$0 \to B_i \hookrightarrow C_i \xrightarrow{\partial_{i-1}} B_{i-1} \to 0$$

is exact. Choose a basis b_i of B_i for $i = -1, \ldots, m$. (Since $B_{-1} = 0$ and $B_m = 0$, b_{-1} and b_m are empty sets.) By the above construction, $b_i b_{i-1}$ is a basis of C_i, which can be compared with the distinguished basis c_i of C_i.

Definition 1.2 *The torsion of C is*

$$\tau(C) = \prod_{i=0}^{m} [b_i b_{i-1} / c_i]^{(-1)^{i+1}} \in \mathbb{F}^*.$$

Lemma 1.3 $\tau(C)$ *does not depend on the choice of b_i.*

Proof. For $i = -1$ and $i = m$ there is nothing to show. So let $i \in \{0, \ldots, m-1\}$. We have to show that

$$[b_i b_{i-1}/c_i]^{(-1)^{i+1}} [b_{i+1} b_i/c_{i+1}]^{(-1)^{i+2}}$$

is independent of the choice of b_i. Indeed, if b_i' is another basis of B_i, then

$$
\begin{aligned}
[b_i' b_{i-1}/c_i] &= [b_i' b_{i-1}/b_i b_{i-1}] \cdot [b_i b_{i-1}/c_i] \\
&= [b_i'/b_i] \cdot [b_i b_{i-1}/c_i],
\end{aligned}
$$

and similarly

$$[b_{i+1} b_i'/c_{i+1}] = [b_i'/b_i] \cdot [b_{i+1} b_i/c_{i+1}].$$

This implies our claim. $\qquad\qquad\qquad\qquad\qquad\qquad\qquad\qquad\qquad\square$

Remarks 1.4

1. The torsion $\tau(C)$ does depend on the distinguished basis c_i of C_i. Indeed, if C' is the same acyclic chain complex C based by $c' = (c_0', \ldots, c_m')$, then

$$\tau(C') = \tau(C) \prod_{i=0}^{m} [c_i/c_i']^{(-1)^{i+1}}.$$

E.g., if $c_i = (c_i^1, c_i^2, c_i^3, \ldots)$ and $c_i' = (c_i^2, c_i^1, c_i^3, \ldots)$, then $\tau(C') = -\tau(C)$. If $c_i' = (\lambda c_i^1, c_i^2, c_i^3, \ldots)$, then $\tau(C') = \lambda^{(-1)^i} \tau(C)$.

2. Let $m = 0$. Then $C = (0 \to C_0 \to 0)$. The acyclicity assumption $H_0(C) = 0$ implies $C_0 = 0$, so $C = 0$. In particular, C is based. Since, by definition, the determinant of an empty matrix is 1, $\tau(C) = 1$.

3. Let $m = 1$. Then $C = (0 \to C_1 \xrightarrow{\partial_0} C_0 \to 0)$ with C_1, C_0 based. C is acyclic iff ∂_0 is an isomorphism. Let c_1 and c_0 be the bases of C_1 and C_0 and let A be the square matrix representing ∂_0 with respect to these bases. The basis b_1 is an empty set, and we choose $b_0 = c_0$. Then

$$\tau(C) = [b_0/c_0]^{-1} [b_1 b_0/c_1] = [b_1 b_0/c_1] = [b_0/c_1] = (\det A)^{-1}.$$

4. If C' is obtained from an acyclic based chain complex C by a shift, i.e., $C_*' = C_{*+1}$, then $\tau(C') = \tau(C)^{-1}$.

5. The torsion $\tau(C)$ defined above is the inverse of the torsion defined by Milnor [24]. $\qquad\qquad\qquad\qquad\qquad\qquad\qquad\qquad\qquad\qquad\qquad\qquad\diamond$

Let $0 \to C' \to C \to C'' \to 0$ be a short exact sequence of chain complexes so that for each i we have a short exact sequence

$$0 \to C'_i \xrightarrow{\alpha_i} C_i \xrightarrow{\beta_i} C''_i \to 0$$

and for each i the diagram

$$
\begin{array}{ccccccccc}
0 & \longrightarrow & C'_i & \xrightarrow{\alpha_i} & C_i & \xrightarrow{\beta_i} & C''_i & \longrightarrow & 0 \\
 & & \downarrow{\scriptstyle \partial'_{i-1}} & & \downarrow{\scriptstyle \partial_{i-1}} & & \downarrow{\scriptstyle \partial''_{i-1}} & & \\
0 & \longrightarrow & C'_{i-1} & \xrightarrow{\alpha_{i-1}} & C_{i-1} & \xrightarrow{\beta_{i-1}} & C''_{i-1} & \longrightarrow & 0
\end{array}
$$

commutes. Assume that C', C, C'' are acyclic and that C'_i, C_i, C''_i have distinguished bases c'_i, c_i, c''_i such that $c_i \sim c'_i c''_i$ for all i .

Theorem 1.5 (Multiplicativity of the torsion) $\tau(C) = \pm \tau(C') \tau(C'')$.

Proof. Let

$$
\begin{aligned}
B_i &= \operatorname{Im}(\partial_i \colon C_{i+1} \to C_i), \\
B'_i &= \operatorname{Im}(\partial'_i \colon C'_{i+1} \to C'_i), \\
B''_i &= \operatorname{Im}(\partial''_i \colon C''_{i+1} \to C''_i).
\end{aligned}
$$

Clearly, $\alpha_i(B'_i) \subset B_i$ and $\beta_i(B_i) \subset B''_i$. We thus have a commutative diagram

$$
\begin{array}{ccccccccc}
0 & \longrightarrow & C'_i & \xrightarrow{\alpha_i} & C_i & \xrightarrow{\beta_i} & C''_i & \longrightarrow & 0 \\
 & & \uparrow & & \uparrow & & \uparrow & & \\
0 & \longrightarrow & B'_i & \xrightarrow{\alpha_i} & B_i & \xrightarrow{\beta_i} & B''_i & \longrightarrow & 0
\end{array}
$$

with injective vertical arrows and exact upper row. We claim that the lower row is exact too. Indeed, exactness at B'_i follows from the injectivity of α_i. The exactness at B''_i follows from a simple diagram chasing and exactness of the first rows. To prove exactness at B_i, let $x_i \in B_i$ with $\beta_i(x_i) = 0$. Exactness of the upper row at C_i shows that there exists $y_i \in C'_i$ with $\alpha_i(y_i) = b_i$. We want to show that $y_i \in B'_i$. By acyclicity of C', this is equivalent to $\partial'_{i-1} y_i = 0$, which, since α_{i-1} is injective, is equivalent to $\alpha_{i-1}(\partial'_{i-1} y_i) = 0$. But this holds true since $\alpha_{i-1}(\partial'_{i-1} y_i) = \partial_{i-1}(\alpha_i(y_i)) = \partial_{i-1} x_i = 0$.

Fix bases b'_i of B'_i and b''_i of B''_i. They combine to a basis $b_i = b'_i b''_i$ of B_i. So,

$$
\begin{aligned}
\tau(C) &= \prod_i [b_i b_{i-1}/c_i]^{(-1)^{i+1}} \\
&= \prod_i [b'_i b''_i b'_{i-1} b''_{i-1}/c_i]^{(-1)^{i+1}} \\
&= \pm \prod_i [b'_i b'_{i-1} b''_i b''_{i-1}/c'_i c''_i]^{(-1)^{i+1}} \\
&= \pm \prod_i \left([b'_i b'_{i-1}/c'_i]^{(-1)^{i+1}} \cdot [b''_i b''_{i-1}/c''_i]^{(-1)^{i+1}} \right) \\
&= \pm \tau(C') \tau(C'').
\end{aligned}
$$

In the fourth equality we used that the transition matrix $(b'_i b'_{i-1} b''_i b''_{i-1}/c'_i c''_i)$ is block-diagonal. $\qquad\qquad\square$

Side remarks on the Euler characteristic 1.6
The Euler characteristic of a chain complex C is the integer

$$
\chi(C) = \sum_{i=0}^{m} (-1)^i \dim C_i.
$$

The multiplicativity of τ should be compared with the obvious additivity of χ with respect to short exact sequences: Given $0 \to C' \hookrightarrow C \twoheadrightarrow C'' \to 0$ we have

$$
\chi(C) = \chi(C') + \chi(C'').
$$

The formula $\dim C_i = \dim H_i(C) + \dim B_i + \dim B_{i-1}$ implies that

$$
\chi(C) = \sum_{i=0}^{m} (-1)^i \dim C_i = \sum_{i=0}^{m} (-1)^i \dim H_i(C). \tag{1.1}
$$

Thus $\chi(C) = 0$ whenever C is acyclic. The torsion is only defined if χ vanishes, i.e., τ is a secondary invariant with respect to χ. $\qquad\qquad\diamond$

We next discuss a duality theorem for torsions. Given an $\mathbb{F}$-vector space D let $D^* = \mathrm{Hom}_{\mathbb{F}}(D, \mathbb{F})$ be its dual. If $d = (d_1, \ldots, d_k)$ is a basis of D, the dual basis $d^* = (d_1^*, d_2^*, \ldots, d_k^*)$ of D^* is defined by

$$
d_i^*(d_j) = \begin{cases} 1 & \text{if } i = j, \\ 0 & \text{if } i \neq j. \end{cases}
$$

Given a homomorphism $h\colon D \to E$ let $h^*\colon E^* \to D^*$ be the dual homomorphism defined by $h^*(y)(x) = y(h(x))$ for any $y \in E^*$, $x \in D$. If d and e are bases of D and E and d^* and e^* are the dual bases of D^* and E^*, then the matrix of h^* with respect to d^* and e^* is the transpose of the matrix of h with respect to d and e.

Definition 1.7 Let

$$C = (0 \to C_m \xrightarrow{\partial_{m-1}} C_{m-1} \to \ldots \to C_1 \xrightarrow{\partial_0} C_0 \to 0)$$

be a chain complex over $\mathbb{F}$. Its *dual chain complex* is

$$C^* = (0 \to C_m^* \xrightarrow{\partial_{m-1}^*} C_{m-1}^* \to \ldots \to C_1^* \xrightarrow{\partial_0^*} C_0^* \to 0),$$

where $C_i^* = C_{m-i}^*$ and $\partial_i^* = (-1)^{i+1}(\partial_{m-i-1})^*$.

Remarks 1.8

1. Since any composable linear homomorphisms g and h obey $(gh)^* = h^*g^*$, C^* is indeed a chain complex.

2. If C_i has a basis c_i, then C_i^* has a basis $c_i^* = c_{m-i}^*$.

3. If C is acyclic, then C^* is acyclic too. $\diamond$

Theorem 1.9 (Duality for torsions) *Let $C = (0 \to C_m \to \ldots \to C_0 \to 0)$ be a based acyclic chain complex over $\mathbb{F}$. Then C^* is a based acyclic chain complex and $\tau(C^*) = \pm\tau(C)^{(-1)^{m+1}}$.*

Proof. Set again $B_i = \operatorname{Im}(\partial_i\colon C_{i+1} \to C_i)$ and set

$$B_{m-i-1}^* = \operatorname{Im}\partial_{m-i-1}^* = \operatorname{Im}(C_{i+1}^* \xleftarrow{\partial_i^*} C_i^*) \subset C_{i+1}^* = C_{m-i-1}^*.$$

Let $a \in B_i$ and $\tilde{a} \in C_{i+1}$ be a lift of a. Let $b \in B_{m-i-1}^* \subset C_{i+1}^*$ and $\tilde{b} \in C_i^*$ be a lift of b. By the definition of ∂_i^*, we have $b(\tilde{a}) = \tilde{b}(a)$, and so $b(\tilde{a})$ and $\tilde{b}(a)$ do not depend on the choices of $\tilde{a}$ and $\tilde{b}$. We thus have a well defined pairing

$$B_i \times B_{m-i-1}^* \to \mathbb{F}, \qquad (a,b) \mapsto b(\tilde{a}) = \tilde{b}(a),$$

which is clearly non-degenerate. The vector space B_i is thus canonically dual to B_{m-i-1}^*. Choose a basis b_i of B_i and let $b_{m-i-1}^* = b_i^*$ be the dual basis of

B^*_{m-i-1}. Then

$$
\begin{aligned}
\tau(C^*) &= \prod_{j=0}^{m} [b^*_j b^*_{j-1}/c^*_j]^{(-1)^{j+1}} \\
&= \prod_{i=0}^{m} [b^*_{m-i} b^*_{m-i-1}/c^*_{m-i}]^{(-1)^{m-i+1}} \\
&= \prod_{i=0}^{m} [b^*_{i-1} b^*_i/c^*_i]^{(-1)^{m-i+1}} \\
&= \pm \prod_{i=0}^{m} [b_i b_{i-1}/c_i]^{(-1)^{m-i}} \\
&= \pm \left(\prod_{i=0}^{m} [b_i b_{i-1}/c_i]^{(-1)^{i+1}} \right)^{(-1)^{m+1}} = \pm\tau(C)^{(-1)^{m+1}}.
\end{aligned}
$$

The second equality follows by the substitution $j = m - i$, and the fourth equality holds since the transition matrix between dual bases a^* and b^* is the inverse transpose of the transition matrix between a and b, and so $[a^*/b^*] = [a/b]^{-1}$.
□

Remarks 1.10
1. The sign $\pm$ in Theorem 1.5 is $(-1)^{\sum_i \dim B'_{i-1} \dim B''_i}$.
2. The sign $\pm$ in Theorem 1.9 is $(-1)^{\sum_i \dim B_{i-1} \dim B_i}$.

2   Computation of the torsion

Throughout this section, fix an acyclic based finite dimensional chain complex

$$
C = (0 \to C_m \xrightarrow{\partial_{m-1}} C_{m-1} \to \ldots \to C_1 \xrightarrow{\partial_0} C_0 \to 0)
$$

over a field $\mathbb{F}$. We briefly discuss three methods computing the torsion $\tau(C) \in \mathbb{F}^*$.

2.1   First method (Matrix τ-chains, [34])

Let the matrix A_i of the chain homomorphism $\partial_i : C_{i+1} \to C_i$, $i = 0, \ldots, m-1$, be given by

$$
A_i = (a^i_{jk})_{\substack{j=1,\ldots,\dim C_{i+1} \\ k=1,\ldots,\dim C_i}}
$$

Notice that the image of a basis vector $c^{i+1}_j \in C_{i+1}$ expressed in the basis vectors $c^i_k \in C_i$ is the j'th row of A_i.

Definition 2.1 A *matrix chain* for C is a collection of sets $\alpha = (\alpha_0, \alpha_1, \ldots, \alpha_m)$, where $\alpha_i \subset \{1, 2, \ldots, \dim C_i\}$, so that $\alpha_0 = \emptyset$. We think of α_i as a set of basis vectors of C_i.

Let $S_i = S_i(\alpha)$ be the submatrix of A_i formed by the entries a^i_{jk} with $j \in \alpha_{i+1}$ and $k \notin \alpha_i$. The situation for $\dim C_{i+1} = 3$, $\dim C_i = 4$ and $\alpha_i = \{1, 4\}$, $\alpha_{i+1} = \{1, 2\}$ is illustrated in Figure 2.1.

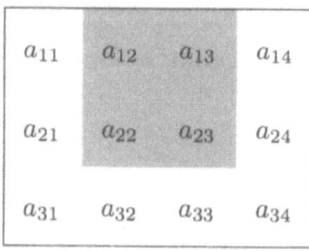

Figure 2.1: A submatrix S_i of A_i.

The matrix chain α is called a τ-*chain* if S_0, S_1, $\ldots$, S_{m-1} are square matrices. The τ-chain α is said to be *non-degenerate* if $\det S_i \neq 0$ for all i.

Theorem 2.2 *Any matrix τ-chain $\alpha = (\alpha_0, \alpha_1, \ldots, \alpha_m)$ for C with $\det S_i(\alpha) \neq 0$ for all even i is non-degenerate, and*

$$\tau(C) = \pm \prod_{i=0}^{m-1} (\det S_i(\alpha))^{(-1)^{i+1}}.$$

Proof. Let $\gamma_i = \operatorname{rk} A_i \geq 0$ be the rank of A_i. By the acyclicity of C, the subcomplex

$$(0 \to \operatorname{Im} \partial_i \hookrightarrow C_i \to C_{i-1} \to \ldots \to C_0 \to 0)$$

is acyclic too, and so its Euler characteristic vanishes. Therefore,

$$\gamma_i = \dim \operatorname{Im} \partial_i = \dim C_i - \dim C_{i-1} + \cdots + (-1)^i \dim C_0. \qquad (2.1)$$

The matrix $S_i = S_i(\alpha)$ has $\#\alpha_{i+1}$ rows and $\dim C_i - \#\alpha_i$ columns. Since S_i is a square matrix, it follows by induction that

$$
\begin{aligned}
\#\alpha_{i+1} &= \dim C_i - \#\alpha_i & (2.2)\\
&= \dim C_i - \dim C_{i-1} + \#\alpha_{i-1} \\
&= \cdots \\
&= \dim C_i - \dim C_{i-1} + \cdots + (-1)^i \dim C_0 \pm \#\alpha_0.
\end{aligned}
$$

But $\#\alpha_0 = 0$, and so, by (2.1), $\#\alpha_{i+1} = \gamma_i$, i.e., S_i is a $(\gamma_i \times \gamma_i)$-matrix.

For even i, S_i is by assumption invertible, and so $\operatorname{rk} S_i = \gamma_i = \operatorname{rk} A_i$, that is, S_i is a maximal non-degenerate submatrix of A_i.

Claim 2.3 *For odd i, S_i is invertible.*

Proof. Let i be odd. Since $\partial_i \circ \partial_{i+1} = 0$, we have $A_{i+1} A_i = 0$. This and $\det S_{i+1} \neq 0$ imply that all rows of A_i are linear combinations of the j-rows of A_i with $j \in \alpha_{i+1}$.

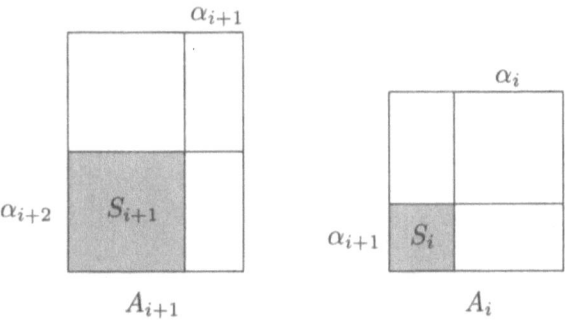

Figure 2.2

Similarly, since $A_i A_{i-1} = 0$ and $\det S_{i-1} \neq 0$, all columns of A_i are linear combinations of the k-columns of A_i with $k \notin \alpha_i$.

These two observations imply that $\operatorname{rk} S_i = \operatorname{rk} A_i = \gamma_i$, and so S_i is invertible. □

After a permutation of the vectors forming the distinguished basis c_i of C_i we may assume that each α_i consists of the last γ_{i-1} vectors in c_i. These permutations do not affect $\tau(C)$ up to sign.

Since S_i has maximal rank, the j-rows of A_i with $j \in \alpha_{i+1}$ form a basis b_i of $B_i = \operatorname{Im}(\partial_i : C_{i+1} \to C_i)$. As a lift $\tilde{b}_i$ of b_i to C_{i+1} we take the sequence of the last γ_i vectors of c_{i+1}. Then

$$[b_i \tilde{b}_{i-1}/c_i] = [b_i \tilde{b}_{i-1}/c_i] = \det \begin{pmatrix} S_i & * \\ 0 & \mathbb{I}_{\#\alpha_i} \end{pmatrix} = \det S_i.$$

This completes the proof of Theorem 2.2. □

Remark 2.4 The sign $\pm$ in Theorem 2.2 is $(-1)^N$, where N is the number of transpositions needed to bring α_i in the form used in the proof. Hence,

$$N = \sum_{i=0}^{m} \# \left\{ (x, y) \mid x < y, \ x \in \alpha_i, \ y \in \{1, 2, \dots, \dim C_i\} \setminus \alpha_i \right\}.$$

For further references we state

Lemma 2.5 *A chain complex over a field is acyclic if and only if it has a non-degenerate τ-chain.*

Proof. Let C be an acyclic chain complex of length m. The proof of Theorem 2.2 shows how to construct a non-degenerate τ-chain for C. We first choose for each A_i with even i a square submatrix S_i of maximal rank $\gamma_i = \mathrm{rk}\, A_i$ computed by (2.1). This determines all the sets α_i, $i = 0, 1, \ldots, m$, and hence also S_i for odd i. Since S_i with i odd has $\#\alpha_{i+1} = \dim C_{i+1} - \gamma_{i+1} = \gamma_i$ rows and $\dim C_i - \#\alpha_{i-1} = \dim C_i - \gamma_{i-1} = \gamma_i$ columns, these S_i are also square matrices. By Theorem 2.2, S_i with odd i are non-degenerate.

Conversely, let $\alpha = (\alpha_0, \alpha_1, \ldots, \alpha_m)$ be a non-degenerate τ-chain for a chain complex C. Formula (2.2) shows that

$$\mathrm{rk}\, A_i \geq \mathrm{rk}\, S_i(\alpha) = \#\alpha_{i+1} = \dim C_i - \dim C_{i-1} + \cdots + (-1)^i \dim C_0.$$

Since C is a chain complex, we have $\mathrm{rk}\, A_{i+1} + \mathrm{rk}\, A_i \leq \dim C_{i+1}$ for all i. These inequalities imply that $\mathrm{rk}\, A_{i+1} + \mathrm{rk}\, A_i = \dim C_{i+1}$ for all i. Now, an easy counting of dimensions shows that C is acyclic. $\square$

2.2 Second method (Chain contractions, [6])

Since C is acyclic, there exists a *chain contraction* $\delta \colon C \to C$, i.e., a sequence of homomorphisms $\delta_i \colon C_i \to C_{i+1}$, $i = 0, \ldots, m$, such that for all i

$$\delta_{i-1}\partial_{i-1} + \partial_i\delta_i = id \colon C_i \to C_i.$$

A chain contraction δ can be obtained as follows.

Set again $B_i = \mathrm{Im}\,(\partial_i \colon C_{i+1} \to C_i)$. Since we work over a field, the short exact sequence

$$0 \to B_i \hookrightarrow C_i \overset{\partial_{i-1}}{\twoheadrightarrow} B_{i-1} \to 0$$

splits as $C_i = B_i \oplus \sigma_i(B_{i-1})$, where $\sigma_i \colon B_{i-1} \to C_i$ is a section of ∂_{i-1}, i.e., $\partial_{i-1} \circ \sigma_i = id$. Define

$$\delta_i \colon C_i = B_i \oplus \sigma_i(B_{i-1}) \to C_{i+1} \quad \text{by} \quad \delta_i(a+b) = \sigma_{i+1}(a), \qquad (2.3)$$

where $a \in B_i$, $b \in \sigma_i(B_{i-1})$. For $b = \sigma_i(b')$ with $b' \in B_{i-1}$ we have

$$
\begin{aligned}
(\delta_{i-1}\partial_{i-1} + \partial_i\delta_i)(a+b) &= \delta_{i-1}(\partial_{i-1}b) + \partial_i(\sigma_{i+1}(a)) \\
&= \delta_{i-1}(b') + a \\
&= \sigma_i(b') + a = b + a = a + b.
\end{aligned}
$$

Thus $\delta_{i-1}\partial_{i-1} + \partial_i\delta_i = id$. Also observe that, by construction, $\delta_i^2 = 0$.

Set

$$C_{even} = \bigoplus_{i\ even} C_i \qquad \text{and} \qquad C_{odd} = \bigoplus_{i\ odd} C_i.$$

Clearly, $\partial + \delta$ maps C_{even} to C_{odd} and C_{odd} to C_{even}.

Theorem 2.6 *For any chain contraction δ,*

$$\tau(C) = \det(\partial + \delta\colon C_{even} \to C_{odd}) = (\det(\partial + \delta\colon C_{odd} \to C_{even}))^{-1}.$$

For a proof, see [6, (16.1)].

Remarks 2.7

1. Since C_{even} and C_{odd} are based and since

$$\dim C_{even} - \dim C_{odd} = \chi(C) = 0,$$

it makes sense to consider the determinants of $\partial + \delta\colon C_{even} \to C_{odd}$ and $\partial + \delta\colon C_{odd} \to C_{even}$. The equality

$$(\partial + \delta)(\partial + \delta) = \partial^2 + \partial\delta + \delta\partial + \delta^2 = id + \delta^2$$

implies that

$$\det(\partial + \delta\colon C_{even} \to C_{odd}) = (\det(\partial + \delta\colon C_{even} \to C_{odd}))^{-1}.$$

Note that δ^2 is a homogeneous operator of degree 2, and so $\det(id + \delta^2) = \det id = 1$.

2. For the chain contraction (2.3), the formula

$$\tau(C) = \det(\partial + \delta\colon C_{even} \to C_{odd})$$

follows at once from the definition of τ. The nontrivial part of the theorem is the proof of this equality if $\delta^2 \neq 0$. $\diamond$

Theorem 2.6 says that the torsion of C is the determinant of a certain matrix associated with C. This is a convenient way to think of torsions.

2.3 Third method (Ray–Singer, [27])

Suppose that $\mathbb{F} = \mathbb{R}$. There is a unique Euclidean metric $\langle \cdot, \cdot \rangle_i$ on C_i such that the distinguished basis c_i is an orthonormal basis. Define the adjoint $\partial_i^\bullet\colon C_i \to C_{i+1}$ by

$$\langle \partial_i^\bullet a, b \rangle_{i+1} = \langle a, \partial_i b \rangle_i, \qquad a \in C_i,\ b \in C_{i+1},$$

and form the Laplacian

$$\Delta_i = \partial_{i-1}^\bullet \partial_{i-1} + \partial_i \partial_i^\bullet\colon C_i \to C_i.$$

The absolute value $|\tau(C)| \in \mathbb{R}_+$ of the torsion can be recovered from the spectra of Δ_i, $i = 0, \ldots, m$, via an explicit formula, see [27].

3   Generalizations and functoriality of the torsion

3.1   Chain complexes with non-vanishing homology

Suppose that $C = (0 \to C_m \to \ldots \to C_0 \to 0)$ is a (not necessarily acyclic) based chain complex over $\mathbb{F}$ and that $H_i(C) = \operatorname{Ker} \partial_{i-1} / \operatorname{Im} \partial_i$ is also based. In this case one can define a torsion $\tau(C) \in \mathbb{F}^*$ as follows. Set

$$B_i = \operatorname{Im} \partial_i \subset C_i, \qquad Z_i = \operatorname{Ker}(\partial_{i-1} : C_i \to C_{i-1}).$$

Then

$$0 \subset B_i \subset Z_i \subset C_i$$

and

$$Z_i / B_i = H_i(C), \qquad C_i / Z_i = B_{i-1}.$$

Let c_i and h_i be the distinguished bases of C_i and $H_i(C)$, respectively. Choose any basis b_i in B_i. Then b_i, h_i, b_{i-1} combine to a basis $b_i h_i b_{i-1}$ of C_i.

Definition 3.1 The *torsion* of C is

$$\tau(C) = \prod_{i=0}^{m} [b_i h_i b_{i-1} / c_i]^{(-1)^{i+1}} \in \mathbb{F}^*.$$

This definition generalizes Definition 1.2. Again, $\tau(C)$ does not depend on the choice of the b_i's, but, of course, depends on the choice of c_i and h_i.

3.2   Chain complexes over associative rings

Let Λ be an associative ring with unit $1 \neq 0$ satisfying the following condition:

(*) For any $r \neq s \in \mathbb{N}$, Λ^r and Λ^s are not isomorphic as Λ-modules.

Here, Λ^r is a direct sum of r copies of Λ. Since Λ is not supposed to be commutative, we cannot define the torsion of a chain complex over Λ by taking determinants. Instead we shall involve the group $K_1(\Lambda)$ defined as follows. For $n \in \mathbb{N}$, let $\operatorname{GL}(n, \Lambda)$ be the group of invertible $(n \times n)$-matrices over Λ. In particular, $\operatorname{GL}(1, \Lambda) = \Lambda^*$ is the group of invertible elements of Λ. Identifying each $M \in \operatorname{GL}(n, \Lambda)$ with the matrix

$$\begin{pmatrix} M & 0 \\ 0 & 1 \end{pmatrix} \in \operatorname{GL}(n+1, \Lambda),$$

we obtain inclusions

$$\operatorname{GL}(1, \Lambda) \subset \operatorname{GL}(2, \Lambda) \subset \ldots.$$

The union $\mathrm{GL}(\Lambda) = \bigcup_{n \geq 0} \mathrm{GL}(n, \Lambda)$ is the *infinite general linear group*. The torsion $\tau(C)$ will be an element of its abelianization

$$K_1(\Lambda) = \mathrm{GL}(\Lambda)/[\mathrm{GL}(\Lambda), \mathrm{GL}(\Lambda)].$$

Consider an acyclic chain complex C of length m over Λ whose chain groups are based free Λ-modules of finite rank. Suppose first that all the modules $B_i = \mathrm{Im}\, \partial_i$ are free. Pick (ordered) bases b_i of B_i and combine them to bases $b_i b_{i-1}$ of C_i. Let c_i be the distinguished basis of C_i. Let $(b_i b_{i-1}/c_i)$ be the transition matrix over Λ defined as in Section 1. Since Λ^r and Λ^s are not isomorphic for $r \neq s$, we have $\#c_i = \#(b_i b_{i-1})$ so that $(b_i b_{i-1}/c_i)$ is a square matrix. Set

$$\hat{\tau}(C) = \prod_{i=0}^{m} (b_i b_{i-1}/c_i)^{(-1)^{i+1}} \in \mathrm{GL}(\Lambda).$$

We see as in the proof of Lemma 1.3 that $\hat{\tau}(C)$ does not depend on the choice of b_i. Define $\tau(C)$ to be the image of $\hat{\tau}(C)$ in $K_1(\Lambda)$.

Even though C_i is free, it may happen that $B_i = \mathrm{Im}\, \partial_i$ is not a free Λ-module, and so one cannot take a basis b_i. To circumvent this problem we use stably free modules. A module B is said to be *stably free* if there exists a free module of finite rank F such that $B \oplus F$ is free.

Lemma 3.2 *The Λ-modules B_i are stably free.*

Proof. Observe that $B_0 = C_0$ is free. Assume by induction that B_{i-1} is stably free. Let F_{i-1} be a free module of finite rank such that $B_{i-1} \oplus F_{i-1}$ is free. Since C is acyclic, we have a short exact sequence

$$0 \to B_i \to C_i \oplus F_{i-1} \xrightarrow{\partial_{i-1} \oplus id} B_{i-1} \oplus F_{i-1} \to 0.$$

Since $B_{i-1} \oplus F_{i-1}$ is free, $\partial_{i-1} \oplus id$ has a section, s_{i-1}. Set $F_i = s_{i-1}(B_{i-1} \oplus F_{i-1})$. Then $C_i \oplus F_{i-1} = B_i \oplus F_i$, i.e., B_i is stably free. $\square$

Given a free based Λ-module F, we denote by $C(F, i)$ the based acyclic free chain complex

$$(\cdots \to 0 \to F \xrightarrow{id} F \to 0 \to \cdots)$$

whose chain modules vanish outside the degrees i and $i + 1$.

Let now C be an acyclic based chain complex of length m over Λ as above. For each $i = 0, 1, \ldots, m-1$ choose a based free Λ-module F_i such that $B_i \oplus F_i$

is free. Observe that the direct sum $C \oplus \bigoplus_i C(F_i, i)$ is an acyclic based chain complex with free modules of boundaries. Define the torsion of C by

$$\tau(C) = \tau\left(C \oplus \bigoplus_{i=0}^{m-1} C(F_i, i)\right) \in K_1(\Lambda).$$

Clearly, $\tau(C)$ does not depend on the choice of bases in $\{F_i\}$. Moreover, $\tau(C)$ is independent of the choice of the stabilizing modules F_i. Indeed, if $\{F_i'\}$ is another family of based free Λ-modules such that $B_i \oplus F_i'$ are free, then

$$\tau\left(C \oplus \bigoplus_i C(F_i', i)\right) = \tau\left(C \oplus \bigoplus_i C(F_i', i) \oplus \bigoplus_i C(F_i, i)\right)$$

$$= \tau\left(C \oplus \bigoplus_i C(F_i, i) \oplus \bigoplus_i C(F_i', i)\right)$$

$$= \tau\left(C \oplus \bigoplus_i C(F_i, i)\right).$$

Here, the first and third equality is obtained by direct application of definitions using the fact that the only non-zero boundary homomorphism in $C(F, i)$ is presented by the identity matrix. The second equality follows from the fact that the chain complexes in question are isomorphic via a basis preserving isomorphism.

The following lemma shows that this concept of torsion generalizes Definition 1.2. First note that for commutative Λ the determinant of matrices defines a surjective group homomorphism $\det \colon K_1(\Lambda) \to \Lambda^*$.

Lemma 3.3 *Let $\mathbb{F}$ be a field. Then the map $\det \colon K_1(\mathbb{F}) \to \mathbb{F}^*$ is an isomorphism of abelian groups. The torsions of chain complexes over $\mathbb{F}$ defined in Sections 1 and 3.2 correspond to each other under this isomorphism.*

Proof. To prove the injectivity of det we need to show that every matrix $A \in \mathrm{GL}(n, \mathbb{F})$ with $\det A = 1$ lies in $[\mathrm{GL}(n, \mathbb{F}), \mathrm{GL}(n, \mathbb{F})]$. Let $E_{ij}(a) \in \mathrm{GL}(n, \mathbb{F})$ be the sum of the identity matrix and the matrix with entry a in the (i, j)'th place and 0 elsewhere. Such a matrix $E_{ij}(a)$ with $i \neq j$ is called *elementary*. The identities

$$(E_{ij}(a))^{-1} = E_{ij}(-a) \quad \text{and} \quad E_{ij}(a)\, E_{jk}(1)\, E_{ij}(-a)\, E_{jk}(-1) = E_{ik}(a)$$

for any distinct i, j, k show that for $n \geq 3$ each elementary matrix in $\mathrm{GL}(n, \mathbb{F})$ is a commutator. Recall now that every square matrix A over a field with

$\det A = 1$ can be reduced to the identity matrix by elementary row operations, i.e., A is a product of elementary matrices. The second claim of the lemma follows from definitions. □

In the next theorem we shall use multiplicative notation for the group operation in $K_1(\Lambda)$. In particular, the neutral element of $K_1(\Lambda)$ represented by the (1×1)-matrix $[1]$ is denoted by 1. The element of $K_1(\Lambda)$ represented by the (1×1)-matrix $[-1]$ is denoted by -1. The definition of the dual chain complex C^* of C parallels Definition 1.7.

Theorem 3.4 (Generalization of Theorems 1.5 and 1.9)

Let $0 \to C' \xrightarrow{\alpha} C \xrightarrow{\beta} C'' \to 0$ be a short exact sequence of free acyclic chain complexes of finite rank over Λ. Assume that C_i', C_i, C_i'' have distinguished bases c_i', c_i, c_i'' such that the image of $(c_i/c_i'c_i'')$ in $K_1(\Lambda)$ is equal to 1 for all i. Then

$$\tau(C) = \pm\tau(C')\,\tau(C'') \in K_1(\Lambda).$$

If $C = (0 \to C_m \to \dots \to C_0 \to 0)$ is a based acyclic chain complex over Λ, then C^* is based acyclic and

$$\tau(C^*) = \pm\tau(C)^{(-1)^{m+1}} \in K_1(\Lambda).$$

Proof. If $B_i = \operatorname{Im} \partial_i$, $B_i' = \operatorname{Im} \partial_i'$, $B_i'' = \operatorname{Im} \partial_i''$, $B_i^* = \operatorname{Im} \partial_i^*$ are free for all i, the claim follows exactly as in the proofs of Theorem 1.5 and 1.9.

If B_i, B_i', B_i'' are not free, let $F' = \oplus_i C(F_i', i)$, $F = \oplus_i C(F_i, i)$ and $F'' = \oplus_i C(F_i'', i)$ be acyclic free chain complexes stabilizing C', C and C'', respectively. We have a short exact sequence

$$0 \to C' \oplus F' \oplus F \xrightarrow{\alpha \oplus id_{F' \oplus F}} C \oplus F' \oplus F \oplus F'' \xrightarrow{\beta \oplus 0 \oplus id_{F''}} C'' \oplus F'' \to 0.$$

Hence,

$$
\begin{aligned}
\tau(C) &= \tau(C \oplus F' \oplus F \oplus F'') \\
&= \pm\tau(C' \oplus F' \oplus F) \cdot \tau(C'' \oplus F'') \\
&= \pm\tau(C') \cdot \tau(C'').
\end{aligned}
$$

If B_i or B_i^* are not free, stabilize C to $C \oplus F$ and stabilize C^* to a certain $C^* \oplus F^\vee$. Since the dual chain complex $(F^\vee)^*$ of $F^\vee$ is the sum of chain complexes of the form $0 \to (F_i^\vee)^* \xrightarrow{\pm id} (F_i^\vee)^* \to 0$, and since the analogous statement holds for F^*, we find

$$
\begin{aligned}
\tau(C) = \tau(C \oplus F) &= \pm\tau(C \oplus F \oplus (F^\vee)^*) \\
&= \pm\tau(C^* \oplus F^* \oplus F^\vee)^{(-1)^{m+1}} \\
&= \pm\tau(C^*)^{(-1)^{m+1}}. \quad\quad\square
\end{aligned}
$$

Exercises 3.5

    1. Extend Definition 3.1 to chain complexes over Λ.

    2. Prove that $K_1(\mathbb{Z}) = \{\pm 1\}$.

3.3 Functoriality of the torsion

Let $\varphi \colon \Lambda \to \Lambda'$ be a ring homomorphism of associative rings with unit. Recall that given a Λ-module K, the tensor product $\Lambda' \otimes_\varphi K$ is the Λ'-module generated by

$$\{\lambda' \otimes k \mid \lambda' \in \Lambda', \, k \in K\}$$

subject to the relations

$$\lambda' \otimes \lambda k = \lambda' \varphi(\lambda) \otimes k \quad (\lambda \in \Lambda, \, \lambda' \in \Lambda', \, k \in K)$$

and the usual bilinearity relations in λ' and k.

    Let C be a based chain complex over Λ. Then $C' = \Lambda' \otimes_\varphi C$ is a based chain complex over Λ'. Indeed, since C_i is free and based, $\Lambda' \otimes_\varphi C_i = \Lambda' \otimes_\varphi \Lambda^r = (\Lambda')^r$ is free and based, where $r = \operatorname{rk} C_i$. If $A_i = (a^i_{jk})$ is the matrix of $\partial_i \colon C_{i+1} \to C_i$, then $A^\varphi_i = (\varphi(a^i_{jk}))$ is the matrix of the boundary homomorphism $C'_{i+1} \to C'_i$.

    The homomorphism φ induces a group homomorphism $\mathrm{GL}(\Lambda) \to \mathrm{GL}(\Lambda')$ by $(a_{ij}) \mapsto (\varphi(a_{ij}))$. Quotienting by the commutator subgroups we obtain a group homomorphism $\varphi_* \colon K_1(\Lambda) \to K_1(\Lambda')$. The next proposition can be easily deduced from definitions and standard homological algebra.

Proposition 3.6 *If C is acyclic, then so is C', and $\tau(C') = \varphi_*(\tau(C))$.*

4 Homological computation of the torsion

4.1 Elementary ideals of modules

Let R be a commutative ring with unit $1 \neq 0$, and let M be a finitely generated R-module. A *presentation* of M is an exact sequence

$$R^m \to R^n \to M \to 0,$$

where $n \in \mathbb{N}$, but m may be infinite. The basis vectors in R^n determine a system of generators in M. Each basis vector in R^m corresponds to a relation between these generators.

    Let A be the $(m \times n)$-matrix of the homomorphism $R^m \to R^n$ with respect to the standard bases in R^m and R^n. This matrix is called a *presentation matrix* of M. Its columns correspond to generators in M, and its rows correspond to linear relations between the generators. Conversely, each $(m \times n)$-matrix over R describes a presentation of an R-module.

Example 4.1 The matrix

$$A = \begin{pmatrix} r_{11} & r_{12} & r_{13} \\ r_{21} & r_{22} & r_{23} \end{pmatrix}$$

presents a module with 3 generators a_1, a_2, a_3, subject to the relations

$$r_{11}a_1 + r_{12}a_2 + r_{13}a_3 = 0, \quad r_{21}a_1 + r_{22}a_2 + r_{23}a_3 = 0.$$

Let A be a presentation $(m \times n)$-matrix for a finitely generated R-module M. For $k \geq 0$, the k-th *elementary ideal* of M is the ideal $E_k(M) = E_k(A) \subset R$ generated by the $(n-k)$-minors (i.e., the $(n-k) \times (n-k)$-subdeterminants) of A. If $n - k \leq 0$, by definition, $E_k(M) = R$. If $n - k > m$, then $E_k(M) = 0$.

Remarks 4.2

1. We may always assume that $m \geq n$. Indeed, if $m < n$, then we can add $n - m$ rows of zeroes to A. This affects neither M nor $E_k(M)$.

2. Some authors call elementary ideals *Fitting ideals* or *determinantal ideals*. ◇

Since the determinant of a matrix is a linear combination of its subdeterminants, we have

$$E_0(M) \subset E_1(M) \subset E_2(M) \subset \cdots . \tag{4.1}$$

Example 4.3 Consider the presentation matrix A of Example 4.1. We have

$$
\begin{aligned}
E_0 &= 0, \\
E_1 &= \left\langle \begin{vmatrix} r_{11} & r_{12} \\ r_{21} & r_{22} \end{vmatrix}, \begin{vmatrix} r_{11} & r_{13} \\ r_{21} & r_{23} \end{vmatrix}, \begin{vmatrix} r_{12} & r_{13} \\ r_{22} & r_{23} \end{vmatrix} \right\rangle, \\
E_2 &= \langle r_{ij} \,|\, i = 1, 2, \ j = 1, 2, 3 \rangle, \\
E_j &= R \ \text{for} \ j \geq 3.
\end{aligned}
$$

Here, we write $\langle r_1, \ldots, r_n \rangle$ for the ideal generated by $r_1, \ldots, r_n$. ◇

Lemma 4.4 *The ideals $E_k(M)$ do not depend on the choice of A.*

Proof. Let A and B be presentation matrices of M and let $a_1, \ldots, a_n$ resp. $b_1, \ldots, b_{n'}$ be the corresponding generators of M. Add the generator b_1 to the set $a_1, \ldots, a_n$. Consider a new presentation of M described by the $(m+1) \times (n+1)$-presentation matrix

$$A' = \begin{bmatrix} * & 1 \\ A & 0 \end{bmatrix}$$

which is obtained from A by adding a relation $b_1 = \sum_{i=1}^{n} r_i a_i$ with $r_i \in R$. We claim that

$$\langle (n-k)\text{-minors of } A \rangle = \langle (n+1-k)\text{-minors of } A' \rangle.$$

The inclusion $\subset$ is clear, and the inclusion $\supset$ follows from the Laplace method of computing determinants.

Applying this argument n' times, we can replace A by a presentation matrix corresponding to the generators $a_1, \ldots, a_n, b_1, \ldots, b_{n'}$. Repeating this argument with $a_1, \ldots, a_n$ replaced by $b_1, \ldots, b_{n'}$, we obtain that $E_k(M)$ does not depend on the choice of generators.

We may hence assume that $n = n'$ and $a_i = b_i$ for all i. Clearly, each row of B is a linear combination of the rows of A. Therefore

$$\langle (n-k)\text{-minors of } A \rangle \;=\; \left\langle (n-k)\text{-minors of } \begin{bmatrix} A \\ B \end{bmatrix} \right\rangle$$

$$=\; \left\langle (n-k)\text{-minors of } \begin{bmatrix} B \\ A \end{bmatrix} \right\rangle$$

$$=\; \langle (n-k)\text{-minors of } B \rangle.$$

This completes the proof of the lemma.   □

4.2   Orders of modules

Recall that a commutative ring R with unit is a *domain* if R has no zero-divisors, i.e., for any $a, b \in R \setminus \{0\}$ we have $ab \neq 0$. An element $a \in R \setminus \{0\}$ is *prime* if $a = bc$ implies that b or c is invertible in R. A domain R is a *unique factorization domain* if every $a \in R \setminus \{0\}$ is a product of prime elements in a unique way (up to ordering and multiplication with invertible elements). Given a subset E of a unique factorization domain, its greatest common divisor $\gcd(E) \in R$ is the generator of the smallest principal ideal containing E. It is defined up to multiplication by invertible elements of R. For instance, if $E = 0$, then $\gcd(E) = 0$, and if $E = R$, then $\gcd(E) = 1$. Polynomial rings and, more generally, Euclidian rings are examples of unique factorization domains.

Let R be a unique factorization domain and let M be a finitely generated R-module. Set

$$\Delta_k(M) = \gcd(E_k(M)) \in R.$$

Thus, $\Delta_k(M)$ is a generator of the smallest principal ideal containing $E_k(M)$. We stress that $\Delta_k(M)$ is defined only up to multiplication with invertible elements of R. The inclusions (4.1) imply that

$$\Delta_{k+1}(M) \,|\, \Delta_k(M) \quad \text{for all } k = 0, 1, \ldots. \tag{4.2}$$

Note that $\Delta_0(M)$ is also denoted by $\mathrm{ord}\, M$ and called the *order* of M.

Remarks 4.5

1. The *rank* $\operatorname{rk} M$ of M is defined to be the dimension of the vector space $\tilde{R} \otimes_R M$ over $\tilde{R}$, the quotient field of R. We have $\Delta_k(M) = 0$ for $k < \operatorname{rk} M$. Indeed, let A be an $(m \times n)$-presentation matrix for M. Then

$$n - k > n - \operatorname{rk} M = n - \dim(\tilde{R} \otimes_R M) = \operatorname{rk} A.$$

Hence all the $(n - k)$-minors of A vanish, so $E_k(M) = 0$ and $\Delta_k(M) = 0$.

2. We have the following equivalences

$$
\begin{aligned}
\operatorname{ord} M = \Delta_0(M) \neq 0 \quad &\Leftrightarrow \quad E_0(M) \neq 0 \\
&\Leftrightarrow \quad A \text{ has a non-vanishing } n\text{-minor} \\
&\Leftrightarrow \quad \tilde{R} \otimes_R M = 0 \\
&\Leftrightarrow \quad \operatorname{rk} M = 0 \\
&\Leftrightarrow \quad M = \operatorname{Tors} M,
\end{aligned}
$$

where A is an $(m \times n)$-presentation matrix of M and

$$\operatorname{Tors} M = \operatorname{Tors}_R M = \{x \in M \mid ax = 0 \text{ for some } a \in R \setminus \{0\}\}$$

is the *torsion submodule* of M.   ◇

Example 4.6 Let $R = \mathbb{Z}$. Then M is a finitely generated abelian group. By the structure theorem for such groups, there are integers $j \geq 0$, $l_1, \ldots, l_k \geq 1$ and prime integers $p_1, \ldots, p_k$ such that

$$M \cong \mathbb{Z}^j \oplus \bigoplus_{i=1}^{k} \left(\mathbb{Z}/p_i^{l_i}\mathbb{Z} \right).$$

A presentation of M

$$\mathbb{Z}^{j+k} \to \mathbb{Z}^{j+k} \twoheadrightarrow M$$

is given by the diagonal matrix

$$A = \operatorname{diag} \left[\underbrace{0, \ldots, 0}_{j \text{ times}}, p_1^{l_1}, \ldots, p_k^{l_k} \right].$$

Thus, $j = \operatorname{rk} M$ and $E_0(M) = (\det A)\,\mathbb{Z}$. So, $\operatorname{ord} M = \Delta_0(M) = \pm \det A$ equals 0 if M is infinite and equals $\pm|M|$ if M is finite.   ◇

Recall that a commutative ring with unit is called *Noetherian* if all its ideals are finitely generated. As is well known, all submodules of a finitely generated module over a Noetherian ring are finitely generated.

Theorem 4.7 [34] *Let R be a Noetherian unique factorization domain. Let $C = (C_m \to \dots \to C_0)$ be a based free chain complex of finite rank over R such that $\operatorname{rk} H_i(C) = 0$ for all i. Let $\tilde{R}$ be the field of fractions of R. Then the based chain complex $\tilde{C} = \tilde{R} \otimes_R C$ is acyclic and*

$$\tau(\tilde{C}) = \prod_{i=0}^{m} (\operatorname{ord} H_i(C))^{(-1)^{i+1}}.$$

Remarks 4.8

1. Since C_i is free, we have a natural inclusion $C_i \subset \tilde{C}_i$. The distinguished basis in C_i yields a distinguished basis in $\tilde{C}_i$, and the boundary homomorphism $\tilde{\partial}_i \colon \tilde{C}_{i+1} \to \tilde{C}_i$ is given by the same matrix as $\partial_i \colon C_{i+1} \to C_i$.

2. A nice feature of Theorem 4.7 is that the homology groups $H_i(C)$ and their orders are often not hard to compute. By Theorem 4.7, they determine the torsion up to multiplication by an invertible element of R. E.g., for $R = \mathbb{Z}$ the torsion is recovered up to sign.

3. For principal ideal domains, Theorem 4.7 was first proven in [25].   ◇

In the proof of Theorem 4.7 we will need the following three lemmas.

Lemma 4.9 *Let M be a finitely generated module over a Noetherian unique factorization domain. Then*

$$\Delta_k(M) = \begin{cases} 0 & \text{if} \quad k < \operatorname{rk} M, \\ \Delta_{k-\operatorname{rk} M}(\operatorname{Tors} M) & \text{if} \quad k \geq \operatorname{rk} M. \end{cases}$$

The case $k < \operatorname{rk} M$ was considered in Remark 4.5.1. For the case $k \geq \operatorname{rk} M$, see for instance [2, Lemma 4.10]. By Remark 4.5.2, $\Delta_{\operatorname{rk}(M)}(M) = \Delta_0(\operatorname{Tors} M) \neq 0$, and so, by (4.2), $\Delta_k(M) \neq 0$ for all $k \geq \operatorname{rk}(M)$.

Given two ideals I and J of R, let $IJ \subset R$ be the ideal additively generated by the elements ab with $a \in I$, $b \in J$.

Lemma 4.10 *For any two ideals I, J of a unique factorization domain R and for any $r \in R$,*

$$\gcd(IJ) = \gcd(I)\gcd(J) \quad \text{and} \quad \gcd(rI) = r\gcd(I).$$

Proof. The equality $\gcd(rI) = r \gcd(I)$ follows from definitions. To prove the equality $\gcd(IJ) = \gcd(I)\gcd(J)$ we divide both I and J by their greatest common divisors. This reduces the claim to the case $\gcd(I) = \gcd(J) = 1$. If a is a prime element of R, then there exist $x \in I$, $y \in J$ non-divisible by a. Then $xy \in IJ$ is not divisible by a. This implies $\gcd(IJ) = 1 = \gcd(I)\gcd(J)$. $\qquad\square$

Lemma 4.11 *Let R be a Noetherian unique factorization domain and let $C = (C_m \to \ldots \to C_0)$ be a based free chain complex of finite rank over R. Denote the matrix of $\partial_i : C_{i+1} \to C_i$ by A_i and the ideal in R generated by the $\operatorname{rk} A_i$-minors of A_i by I_i. Then $\operatorname{ord}\operatorname{Tors} H_i(C) = \gcd(I_i)$ for all $i = 0, 1, \ldots, m$.*

Proof. Consider the exact sequence

$$C_{i+1} \xrightarrow{\partial_i} C_i \twoheadrightarrow \operatorname{Coker}\partial_i \to 0.$$

The matrix A_i is a presentation matrix for $\operatorname{Coker}\partial_i$. By definition, I_i is the first non-zero ideal in the sequence

$$E_0(\operatorname{Coker}\partial_i) \subset E_1(\operatorname{Coker}\partial_i) \subset \ldots,$$

and so $\gcd(I_i)$ is the first non-zero element in the sequence $\Delta_k(\operatorname{Coker}\partial_i)$, where $k = 0, 1, 2, \ldots$. By Lemma 4.9, $\gcd(I_i) = \Delta_0(\operatorname{Tors}\operatorname{Coker}\partial_i)$.

Since C_{i-1} is a free R-module, $\operatorname{Tors} C_{i-1} = 0$. The exactness of

$$0 \to H_i(C) \hookrightarrow \operatorname{Coker}\partial_i \xrightarrow{\partial_{i-1}} C_{i-1} \to 0$$

implies $\operatorname{Tors}\operatorname{Coker}\partial_i = \operatorname{Tors} H_i(C)$. Therefore

$$\gcd(I_i) = \Delta_0(\operatorname{Tors}\operatorname{Coker}\partial_i) = \operatorname{ord}\operatorname{Tors} H_i(C). \qquad\square$$

Proof of Theorem 4.7. We first check that $\tilde{C}$ is acyclic. Let $x \in \operatorname{Ker}(\tilde{\partial}_i : \tilde{C}_{i+1} \to \tilde{C}_i)$. There exists $a \in R \setminus \{0\}$ such that $ax \in C_{i+1}$. We have $\partial_i(ax) = a\partial_i(x) = 0$. Since $H_i(C) = \operatorname{Tors} H_i(C)$, there is $a' \in R \setminus \{0\}$ such that the homology class of the cycle $a'ax$ is zero. Thus, there exists $y \in C_{i+2}$ with $a'ax = \partial_{i+1}(y)$. Since R is a domain, $a'a \neq 0$, and so $x = \tilde{\partial}_{i+1}((a'a)^{-1}y)$. This proves the acyclicity of $\tilde{C}$.

Let A_i be the matrix of $\partial_i : C_{i+1} \to C_i$ and I_i be the ideal in R generated by the $\operatorname{rk} A_i$-minors of A_i. The ideal $\prod_{i \text{ even}} I_i \subset R$ is generated by the products $\prod_{i \text{ even}} b_i$, $b_i \in I_i$. We claim that

$$\tau(\tilde{C}) \cdot \prod_{i \text{ even}} I_i = \prod_{i \text{ odd}} I_i. \tag{4.3}$$

Writing $\tau(\tilde{C}) = \frac{x}{y}$, $x, y \in R \setminus \{0\}$, we can rewrite (4.3) as

$$x \prod_{i \text{ even}} I_i = y \prod_{i \text{ odd}} I_i. \tag{4.4}$$

Choose for each even i a minor S_i of A_i of rank $\operatorname{rk} A_i$ such that $\det S_i \neq 0$. As we saw in the proof of Theorem 2.2 and Lemma 2.5, this determines a matrix τ-chain for C and hence submatrices S_i of A_i for odd i. By Theorem 2.2,

$$\tau(\tilde{C}) \cdot \prod_{i \text{ even}} \det S_i = \pm \prod_{i \text{ odd}} \det S_i \in \prod_{i \text{ odd}} I_i.$$

Since, for each i, the $\operatorname{rk} A_i$-minors of A_i generate I_i,

$$\tau(\tilde{C}) \cdot \prod_{i \text{ even}} I_i \subset \prod_{i \text{ odd}} I_i.$$

Applying the same argument to the shifted complex $C'_* = C_{*+1}$ we obtain

$$\tau(\tilde{C})^{-1} \cdot \prod_{i \text{ odd}} I_i \subset \prod_{i \text{ even}} I_i$$

(cf. Remark 1.4.4). These two inclusions imply (4.3) and (4.4).

Taking the greatest common divisor on both sides of (4.4) and using Lemmas 4.10 and 4.11, we obtain

$$x \prod_{i \text{ even}} \operatorname{ord} \operatorname{Tors} H_i(C) = y \prod_{i \text{ odd}} \operatorname{ord} \operatorname{Tors} H_i(C),$$

or, equivalently, $\tau(\tilde{C}) = \prod_{i=0}^{m} (\operatorname{ord} \operatorname{Tors} H_i(C))^{(-1)^{i+1}}$. Observe now that the assumption $\operatorname{rk} H_i(C) = 0$ implies $\operatorname{Tors} H_i(C) = H_i(C)$ for all i. This completes the proof of Theorem 4.7. $\qquad\square$

Chapter II
Topological Theory of Torsions

5 Basics of algebraic topology

In this section, we collect some basic concepts and results from algebraic topology. We refer to [21] and [32] for details and further information.

5.1 The fundamental group of a topological space

Let X be a space with base point x. A *loop* in (X, x) is a continuous map $\alpha\colon I = [0,1] \to X$ with $\alpha(0) = \alpha(1) = x$. Given two loops α and β in X, there product $\alpha\beta$ is defined by

$$(\alpha\beta)(t) = \left\{ \begin{array}{ll} \alpha(2t), & 0 \le t \le \frac{1}{2}, \\ \beta(2t-1), & \frac{1}{2} \le t \le 1. \end{array} \right. \tag{5.1}$$

The inverse α^{-1} of a loop α is the loop $\alpha^{-1}(t) = \alpha(1-t)$, and the constant loop is the loop $I \to x \in X$.

A *homotopy of loops* is a continuous map $F\colon I \times I \to X$ such that for each $s \in I$ the map $F(\cdot, s)$ is a loop in (X, x). Given a loop α in X, we write $[\alpha]$ for the class of loops in (X, x) homotopic to α. Under the product induced by (5.1), the homotopy classes of loops in (X, x) form a group called the *fundamental group* of X and denoted by $\pi_1(X, x)$.

Let x' be a point in the path-component of x. Choose a path $\gamma\colon I \to X$ with $\gamma(0) = x$, $\gamma(1) = x'$. The induced map

$$\gamma_\# \colon \pi_1(X, x) \to \pi_1(X, x'), \quad [\alpha] \mapsto [\gamma^{-1}\alpha\gamma]$$

is a group isomorphism. This isomorphism may depend on the choice of γ.

Let (Y, y) be another pointed topological space and let $f\colon (X, x) \to (Y, y)$ be a continuous map. Then f induces a group homomorphism

$$f_\# \colon \pi_1(X, x) \to \pi_1(Y, y), \quad [\alpha] \mapsto [f \circ \alpha].$$

Thus, π_1 is a functor from the category of pointed spaces to the category of groups.

A topological space X is *simply connected* if X is path-connected and $\pi_1(X, x) = 1$ for $x \in X$. For instance, Euclidean space $\mathbb{R}^n$ with any n and the n-dimensional sphere S^n with $n \geq 2$ are simply connected.

5.2 Covering spaces

A surjective continuous map $p\colon \widetilde{X} \to X$ is called a *covering projection* if each point $x \in X$ has an open neighbourhood $U \subset X$ such that $p^{-1}(U)$ is a union of disjoint open subsets of $\widetilde{X}$ each of which is mapped homeomorphically onto U by p. The space $\widetilde{X}$ is then called a *covering space* of X, and X is called the *base* of p. If $\widetilde{X}$ is simply connected, $p\colon \widetilde{X} \to X$ is called a *universal covering* of X. Two coverings $p_1\colon \widetilde{X}_1 \to X$ and $p_2\colon \widetilde{X}_2 \to X$ are *equivalent* if there is a homeomorphism $f\colon \widetilde{X}_1 \to \widetilde{X}_2$ such that $p_1 = p_2 \circ f$.

A space is said to be *locally path-connected* if each of its points has a basic family of path-connected neighbourhoods. Observe that connected locally path-connected spaces are path-connected.

Lemma 5.1 (Unique Lifting Property) *Let $p\colon \widetilde{X} \to X$ be a covering, let Z be a connected locally path-connected space and let $f\colon Z \to X$ be a continuous map. Fix $z \in Z$ and $\tilde{x} \in p^{-1}(f(z))$. If $f_{\#}\pi_1(Z, z) \subset p_{\#}\pi_1(\widetilde{X}, \tilde{x})$, then there exists a unique continuous map $\tilde{f}\colon Z \to \widetilde{X}$ such that $\tilde{f}(z) = \tilde{x}$ and $p \circ \tilde{f} = f$:*

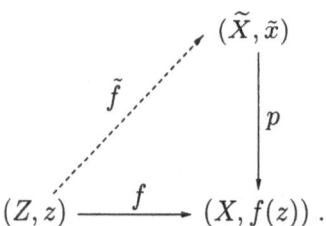

For a proof we refer to [21, Chapter V, Lemma 3.2 and Theorem 5.1].

Corollary 5.2 *If $p\colon \widetilde{X} \to X$ is a covering with connected $\widetilde{X}$ and simply connected locally path-connected X, then p is a homeomorphism.*

Proof. It suffices to apply Lemma 5.1 to $Z = X$ and the identity map $f = \mathrm{id}\colon X \to X$. □

Corollary 5.3 *Let X be a connected locally path-connected space. Assume that $p_1\colon (\widetilde{X}_1, \tilde{x}_1) \to (X, x)$ and $p_2\colon (\widetilde{X}_2, \tilde{x}_2) \to (X, x)$ are coverings of X such that $\widetilde{X}_1, \widetilde{X}_2$ are connected and $(p_1)_{\#}(\pi_1(\widetilde{X}_1, \tilde{x}_1)) = (p_2)_{\#}(\pi_1(\widetilde{X}_2, \tilde{x}_2)) \subset \pi_1(X, x)$. Then these two coverings are equivalent.*

Proof. Since X is locally path-connected, so are $\widetilde{X}_1$ and $\widetilde{X}_2$. The existence part of the Unique Lifting Property implies that there are maps $f\colon (\widetilde{X}_1, \tilde{x}_1) \to (\widetilde{X}_2, \tilde{x}_2)$ and $g\colon (\widetilde{X}_2, \tilde{x}_2) \to (\widetilde{X}_1, \tilde{x}_1)$ such that $p_2 \circ f = p_1$ and $p_1 \circ g = p_2$. The uniqueness part implies that $g \circ f = id_{\widetilde{X}_1}$ and $f \circ g = id_{\widetilde{X}_2}$, and so f and g are mutually inverse homeomorphisms. $\square$

A space X is said to be *semilocally simply connected* if each point $x \in X$ has a neighbourhood U such that the homomorphism $\pi_1(U, x) \to \pi_1(X, x)$ is trivial. For a proof of the following theorem, we refer to [21, Chapter V, Theorem 10.2].

Theorem 5.4 *Let X be a connected, locally path-connected and semilocally simply connected space. Then there exists a universal covering of X.*

Lemma 5.1 implies that under the conditions of Theorem 5.4 the universal covering of X is unique up to equivalence of coverings.

5.3 Group actions

Let G be a group of homeomorphisms of a topological space Z. Then G is said to be *properly discontinuous* if for all $z \in Z$ there is an open neighbourhood V of z such that the sets gV, $g \in G$, are pairwise disjoint. In particular, G has no fixed points, i.e., G acts freely. Conversely, if G is finite and fixed-point free, then it is properly discontinuous.

Let $p\colon Z \to Z/G$ be the projection onto the set of equivalence classes of a G-action on Z. We endow Z/G with the quotient topology, that is, a set $U \subset Z/G$ is open if $p^{-1}(U)$ is open in Z. The following lemma is obvious.

Lemma 5.5 *If G is a properly discontinuous group of homeomorphisms of Z, then $p\colon Z \to Z/G$ is a covering projection.*

Theorem 5.6 *If Z is simply connected, then $\pi_1(Z/G) = G$.*

Sketch of a proof. Fix $z \in Z$. Since Z is path-connected, for any $g \in G$ there is a path $\gamma_g\colon I \to Z$ with $\gamma_g(0) = z$ and $\gamma_g(1) = gz$. Clearly, $p \circ \gamma_g$ is a loop in Z/G. Since Z is simply connected, the formula $g \mapsto [p \circ \gamma_g]$ yields a well defined map $G \to \pi_1(Z/G, p(z))$. That this map is a homomorphism is clear, and that it is both injective and surjective follows from Lemma 5.1. $\square$

Corollary 5.7 *Let $p\colon \widetilde{X} \to X$ be a universal covering of a connected locally path-connected space X. Then the group $\pi_1(X)$ acts as a properly discontinuous group of homeomorphisms of $\widetilde{X}$ and $\widetilde{X}/\pi_1(X) = X$.*

Proof. A *covering transformation* of $p\colon \widetilde{X} \to X$ is a homeomorphism $h\colon \widetilde{X} \to \widetilde{X}$ such that $p \circ h = p$. The covering transformations form a group under composition. It is denoted by $\mathrm{Aut}(p)$. By the Unique Lifting Property, for any two points $x, y \in \widetilde{X}$ with $p(x) = p(y)$ there is a (unique) covering transformation of p sending x to y. Therefore $\widetilde{X}/\mathrm{Aut}(p) = X$. It follows from the definition of a covering that the action of $\mathrm{Aut}(p)$ on $\widetilde{X}$ is properly discontinuous. By Theorem 5.6, $\pi_1(X) = \mathrm{Aut}(p)$. $\square$

Examples 5.8

1. The infinite cyclic group $\{T^n\}_{n\in\mathbb{Z}}$ acts on the real line $\mathbb{R}$ by $T(x) = x+1$ for $x \in \mathbb{R}$. This action is properly discontinuous. Clearly, $\mathbb{R}/\{T^n\}_{n\in\mathbb{Z}} = S^1$. By Lemma 5.5 and Theorem 5.6, the projection $p\colon \mathbb{R} \to S^1$, $x \mapsto e^{2\pi i x}$ is a universal covering, and $\pi_1(S^1) = \{T^n\}_{n\in\mathbb{Z}}$.

2. (Lens spaces) Let $n \in \mathbb{N}$ and

$$S^{2n-1} = \left\{ (z_1, \ldots, z_n) \in \mathbb{C}^n \mid \sum_{i=1}^{n} |z_i|^2 = 1 \right\}.$$

Pick an integer $p \geq 2$ (not necessarily prime) and let $q_1, q_2, \ldots, q_n$ be integers relatively prime to p. The cyclic group $G_p = \{\zeta \in \mathbb{C} \mid \zeta^p = 1\}$ acts on S^{2n-1} by

$$\zeta(z_1, z_2, \ldots, z_n) = (\zeta^{q_1} z_1, \zeta^{q_2} z_2, \ldots, \zeta^{q_n} z_n).$$

This action is fixed-point free and therefore properly discontinuous. The orbit space S^{2n-1}/G_p is called a *lens space* and is denoted $L(p; q_1, q_2, \ldots, q_n)$. For $n \geq 2$, S^{2n-1} is simply connected, so $\pi_1(L(p; q_1, \ldots, q_n)) = G_p$. In particular, the lens space $L(2; q_1, q_2, \ldots, q_n) = L(2; 1, 1, \ldots, 1)$ is nothing but the $(2n-1)$-dimensional real projective space $\mathbb{R}P^{2n-1}$.

5.4   Maximal abelian and maximal free abelian coverings

Consider a connected locally path-connected space X. Let $\widetilde{X} \to X$ be a universal covering of X. Since $\pi = \pi_1(X)$ acts properly discontinuously on $\widetilde{X}$, so does the commutator subgroup $[\pi, \pi] \subset \pi$. Set $\widehat{X} = \widetilde{X}/[\pi, \pi]$. The covering $\widetilde{X} \to X$ induces a covering $\hat{p}\colon \widehat{X} \to X$. Clearly, $\hat{p}_\#(\pi_1(\widehat{X})) = [\pi, \pi]$, and by Corollary 5.3 this determines the covering $\hat{p}$. It is called the *maximal abelian covering* of X. The group $H = \pi/[\pi, \pi] = H_1(X; \mathbb{Z})$ acts on $\widehat{X}$ properly discontinuously and $\widehat{X}/H = X$.

Set $G = H/\operatorname{Tors} H$, where $\operatorname{Tors} H$ is the subgroup of H consisting of elements of finite order. Set $\bar{X} = \widehat{X}/\operatorname{Tors} H$. The covering $\hat{p}\colon \widehat{X} \to X$ induces a covering $\bar{X} \to X$ called the *maximal free abelian covering* of X. The group G acts on $\bar{X}$ properly discontinuously and $\bar{X}/G = X$.

We obtain a tower of covering spaces $\widetilde{X} \to \widehat{X} \to \bar{X} \to X$, where $X = \widetilde{X}/\pi = \widehat{X}/H = \bar{X}/G$.

Examples 5.9
 1. If $X = S^1$, then $\bar{X} = \widehat{X} = \widetilde{X} = \mathbb{R}$.
 2. If $X = L(p; q_1, \ldots, q_n)$, then $\bar{X} = X$ and $\widehat{X} = \widetilde{X} = S^{2n-1}$.
 3. If $X = S^1 \times L(p; q_1, \ldots, q_n)$, then $\bar{X} = \mathbb{R} \times L(p; q_1, \ldots, q_n)$ and $\widehat{X} = \widetilde{X} = \mathbb{R} \times S^{2n-1}$.

Remark 5.10 All (topological) manifolds are locally path-connected and semi-locally simply connected. Therefore, all the definitions and results of Sections 5.2 – 5.4 apply to connected manifolds.

5.5 Orientability

Let M be a connected topological manifold of dimension m and let $\gamma \colon S^1 \hookrightarrow M$ be an embedding of the circle in M. A regular neighbourhood of S^1 in M is homeomorphic either to the solid torus $S^1 \times B$ where $B = B^{m-1}$ is an $(m-1)$-dimensional ball, or to the twisted solid torus obtained from $[0,1] \times B$ by the identification $(0, b) = (1, j(b))$ where $j \colon B \to B$ is an orientation reversing involution and b runs over B. We set $w_1(\gamma) = 0$ in the first case and $w_1(\gamma) = 1$ in the second case. It is easy to see that $w_1(\gamma)$ does not depend on the choice of γ in its homotopy class. We thus obtain a homomorphism $w_1 \colon \pi_1(M) \to \mathbb{Z}/2\mathbb{Z}$ called the *first Stiefel–Whitney class* of M. Observe that $w_1(\alpha) = w_1(\alpha^{-1})$ for all $\alpha \in \pi_1(M)$, and that $w_1 = 0$ iff M is orientable. The induced homomorphism $H_1(M) \to \mathbb{Z}/2\mathbb{Z}$ is also denoted by w_1.

Lemma 5.11 *Let $\widehat{M}$ be the maximal abelian covering of a connected topological manifold M. Then $\widehat{M}$ is orientable.*

Proof. Set $\pi = \pi_1(M)$ and $\hat{\pi} = \pi_1(\widehat{M})$. Let $\hat{p} \colon \widehat{M} \to M$ be the covering projection. It follows from definitions that the diagram

$$
\begin{array}{ccccc}
\hat{\pi} & \longrightarrow & H_1(\widehat{M}) & \xrightarrow{\;w_1(\widehat{M})\;} & \mathbb{Z}/2\mathbb{Z} \\[2mm]
\Big\downarrow{\scriptstyle \hat{p}_\#} & & \Big\downarrow{\scriptstyle \hat{p}_*} & & \Big\| \\[2mm]
\pi & \longrightarrow & H_1(M) & \xrightarrow{\;w_1(M)\;} & \mathbb{Z}/2\mathbb{Z}
\end{array}
$$

commutes. The identities $\hat{p}_\#(\hat{\pi}) = [\pi, \pi]$ and $H_1(M) = \pi/[\pi, \pi]$ now imply that $w_1(\widehat{M}) = 0$. Therefore, $\widehat{M}$ is orientable. $\square$

5.6   CW-complexes

Let $D^k = \{x \in \mathbb{R}^k \,|\, |x| \leq 1\}$ be the closed k-ball and let $\partial D^k = S^{k-1}$ be its boundary. Let $X \subset X'$ be a pair of topological spaces. We say that X' is obtained from X by *adjoining k-cells* if there exists a continuous map $f = \coprod_i f_i \colon \coprod_i D^k \to X'$ such that

- $f|_{\coprod_i \text{Int } D^k}$ is a homeomorphism onto $X' \setminus X$,

- a subset U of X' is open if and only if $U \cap X$ is open in X and $f_i^{-1}(U)$ is open in D^k for all i.

Here i runs over a certain (possibly infinite) set of indices. Each map f_i is called a *characteristic map*, and $e_i^k = f_i(\text{Int } D^k)$ is called an *open k-cell*. We also set $\bar{e}_i^k = f_i(D^k)$. Note that $\bar{e}_i^k$ does not need to be homeomorphic to D^k (cf. Figure 5.1).

Definition 5.12 A CW-complex is a Hausdorff space X along with an increasing sequence of closed subspaces $X^0 \subset X^1 \subset X^2 \subset \ldots$ such that

(a) X^0 is a discrete space,

(b) $X = \bigcup_{k \geq 0} X^k$,

(c) for $k \geq 1$, X^k is obtained from X^{k-1} by adjoining k-cells,

(d) a subset U of X is open if and only if $U \cap X^k$ is open in X^k for each k.

The subset X^k is called the *k-skeleton* of X. A CW-complex X is said to be *finite* if it is formed by a finite number of cells. It is known that a CW-complex is finite if and only if its underlying topological space is compact.

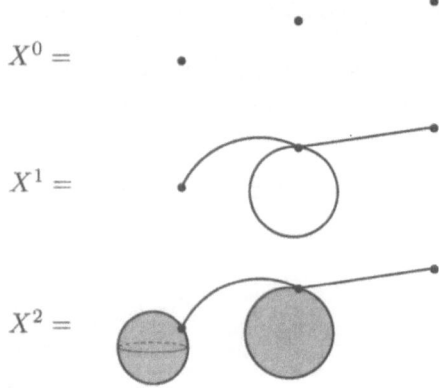

Figure 5.1: A 2-dimensional CW-complex.

Let X and Y be CW-complexes. A continuous map $f\colon X \to Y$ is *cellular* if $f(X^k) \subset Y^k$ for all $k \geq 0$. Any continuous map between CW-complexes is homotopic to a cellular map.

Let X and Y be CW-complexes, and let $\{e_i\}$ be the cells of X and $\{d_j\}$ be the cells of Y. Then the cells $\{e_i \times d_j\}$ form a CW-decomposition of $X \times Y$. If $f\colon D^m \to X$ and $g\colon D^n \to Y$ are characteristic maps of e and d, respectively, then

$$f \times g\colon D^{m+n} = D^m \times D^n \to X \times Y$$

is a characteristic map for $e \times d$. The k-skeleton of $X \times Y$ is

$$(X \times Y)^k = (X^0 \times Y^k) \cup (X^1 \times Y^{k-1}) \cup \cdots \cup (X^k \times Y^0).$$

Notice that CW-complexes are locally path-connected and semilocally simply connected. Therefore, any connected CW-complex X is path-connected and admits a unique universal covering $p\colon \tilde{X} \to X$.

Lemma 5.13 *The structure of a CW-complex on X induces the structure of a CW-complex on $\tilde{X}$. The group $\mathrm{Aut}(p) = \pi_1(X)$ acts freely and transitively on the set of cells of $\tilde{X}$ lying over any cell of X.*

Proof. For $k \geq 0$, set $\tilde{X}^k = p^{-1}(X^k) \subset \tilde{X}$. The sequence $\tilde{X}^0 \subset \tilde{X}^1 \subset \ldots$ satisfies all the conditions of Definition 5.12. The k-cells of $\tilde{X}$ are obtained from the k-cells of X as follows. Let $e \subset X$ be an open k-cell with characteristic map $f\colon D^k \to X^k$. Fix $d \in \mathrm{Int}\, D^k$. By the Unique Lifting Property, for any point $\tilde{d} \in p^{-1}(f(d))$ there is a unique lift $\tilde{f}\colon D^k \to \tilde{X}$ of f such that $\tilde{f}(d) = \tilde{d}$. Then $\tilde{f}(\mathrm{Int}\, D^k)$ is an open k-cell of $\tilde{X}$ with characteristic map $\tilde{f}$. The covering projection p maps this cell homeomorphically onto e. If $h \in \mathrm{Aut}(p)$ then $h \circ \tilde{f}$ is another lift of f. The group $\mathrm{Aut}(p)$ thus acts on the set of lifts of f. Since $\mathrm{Aut}(p)$ acts freely and transitively on $p^{-1}(f(d))$, the action of $\mathrm{Aut}(p)$ on the lifts of f is free and transitive. $\qquad\square$

5.7 The cellular chain complex

Let X be a CW-complex whose (open) cells are all oriented. We define the *cellular chain complex* of X as follows. We begin with a preliminary definition. Given an open k-cell e_i^k of X, consider its characteristic map $f_i^k\colon D^k \to X$ and consider the composition of maps

$$f_{i,j}^k\colon S^{k-1} = \partial D^k \to X^{k-1} \to X^{k-1}/X^{k-2} = \bigvee_j S_j^{k-1} \to S_j^{k-1}$$

where the first map is the restriction of f_i^k to ∂D^k and the second map is the projection collapsing X^{k-2} to a point x. The space X^{k-1}/X^{k-2} is a bunch of spheres $\{S_j^{k-1}\}_j$ numerated by the $(k-1)$-cells $\{e_j^{k-1}\}_j$ of X. The last map further collapses all but one of the spheres. The orientations of e_i^k and e_j^{k-1} induce orientations of $S^{k-1} = \partial D^k$ and S_j^{k-1}. (We use the "outward vector first" convention for the boundary which means that the outward looking vector on ∂D^k followed by a positive basis of tangent vectors of S^{k-1} should give a positive basis of tangent vectors of D^k.) Now, the degree $\deg f_{i,j}^k \in \mathbb{Z}$ of $f_{i,j}^k$ is defined. For fixed i and k, only finitely many of the numbers $\deg f_{i,j}^k$ do not vanish. Indeed, let $g_i^k \colon S^{k-1} \to \bigvee_j S_j^{k-1}$ be the composition of the first two maps above. For each j, let $U_j = S_j^{k-1} \setminus \{x\}$, and let $V \subset \bigvee_j S_j^{k-1}$ be a small open neighbourhood of x. Clearly, $\{U_j\}_j$, V is an open cover of $\bigvee_j S_j^{k-1}$. Since $g_i^k(S^{k-1})$ is a compact subset of $\bigvee_j S_j^{k-1}$, it is contained in the union of V with finitely many U_j. This implies that $\deg f_{i,j}^k$ vanishes for all but finitely many j.

Define the i-th cellular chain group of X by $C_k(X) = \oplus_i \mathbb{Z} e_i^k$. Thus, $C_k(X)$ is the free abelian group generated by the set of oriented k-cells of X. Define the boundary homomorphism $\partial \colon C_k(X) \to C_{k-1}(X)$ by

$$\partial e_i^k = \sum_j \left(\deg f_{i,j}^k \right) e_j^{k-1}.$$

As is well known,

$$C(X) = (\ldots \to C_k(X) \xrightarrow{\partial} C_{k-1}(X) \to \ldots)$$

is a chain complex, and its homology $H_*(C(X))$ equals the singular homology of X with integer coefficients. In particular, $H_*(C(X))$ depends neither on the choice of orientations of the cells of X nor on the CW-structure on X.

If X is a finite CW-complex, then the *Euler characteristic* of X

$$\chi(X) = \sum_k (-1)^k \operatorname{rk}_{\mathbb{Z}} C_k(X) = \sum_k (-1)^k \dim H_k(X; \mathbb{R})$$

does not depend on the CW-structure on X. Note that $\operatorname{rk}_{\mathbb{Z}} C_k(X)$ is just the number of k-cells of X.

6 The Reidemeister–Franz torsion

Given a group π, denote by $\mathbb{Z}[\pi]$ the group ring of π with integer coefficients. As an additive group, $\mathbb{Z}[\pi]$ is free abelian with basis π, and multiplication in $\mathbb{Z}[\pi]$ is induced by the one in π.

6.1 The Reidemeister–Franz torsion of a CW-complex

Let X be a finite connected CW-complex and let $p \colon \widetilde{X} \to X$ be its universal covering with CW-structure induced from X. We orient all open cells of X and orient the cells of $\widetilde{X}$ in such a way that the restriction of p to each cell is orientation preserving. Fix $x \in X$ and set $\pi = \pi_1(X, x)$. The action of π on $\widetilde{X}$ by covering transformations induces an action of π on the cellular chain groups $C_k(\widetilde{X})$. Extend this action linearly to an action of $\mathbb{Z}[\pi]$. In this way, $C_k(\widetilde{X})$ becomes a $\mathbb{Z}[\pi]$-module. Note that the boundary homomorphism $\partial \colon C_k(\widetilde{X}) \to C_{k-1}(\widetilde{X})$ is linear over $\mathbb{Z}[\pi]$ for all $k \geq 1$.

Let $\{e_i^k\}$ be the set of oriented k-cells of X ordered in an arbitrary way. Choose over each e_i^k a k-cell $\tilde{e}_i^k$ in $\widetilde{X}$. Then the set $\{\tilde{e}_i^k\}$ is a basis of $C_k(\widetilde{X})$, i.e.,

$$C_k(\widetilde{X}) = \bigoplus_i \mathbb{Z}[\pi]\, \tilde{e}_i^k.$$

Thus, $C(\widetilde{X})$ is a free and based chain complex over $\mathbb{Z}[\pi]$. However, this chain complex is not acyclic. Indeed, already its 0-dimensional homology is non-trivial: $H_0(C(\widetilde{X})) = H_0(\widetilde{X}; \mathbb{Z}) = \mathbb{Z}$. In order to produce acyclic chain complexes, we proceed as follows.

Let Λ be an associative ring with unit such that for any $r \neq s \in \mathbb{N}$, Λ^r and Λ^s are not isomorphic as Λ-modules. Given a ring homomorphism $\varphi \colon \mathbb{Z}[\pi] \to \Lambda$, consider the chain complex

$$C^\varphi(X) = \Lambda \otimes_\varphi C(\widetilde{X}).$$

Then $C_k^\varphi(X) = \oplus_i \Lambda \tilde{e}_i^k$, and the matrix of the boundary homomorphism

$$\partial \colon C_k^\varphi(X) \to C_{k-1}^\varphi(X)$$

is obtained by applying φ to the matrix of $\partial \colon C_k(\widetilde{X}) \to C_{k-1}(\widetilde{X})$. The homology of $C^\varphi(X)$ is denoted by $H_*^\varphi(X)$. It is called the *twisted homology* of X. The groups $H_*^\varphi(X)$ are isomorphic to the twisted singular homology of X with coefficients in Λ. In particular, they depend neither on the CW-structure on X nor on the choice of cell orientations.

Assume that the chain complex $C^\varphi(X)$ is acyclic, i.e.,

$$H_*(C^\varphi(X)) = H_*^\varphi(X) = 0.$$

Under this condition we set

$$\tau_\varphi(X, \tilde{e}) = \tau(C^\varphi(X)) \in K_1(\Lambda),$$

where $\tilde{e}$ stands for the basis of $C^\varphi(X)$ determined by the oriented ordered cells $\{\tilde{e}_i^k\}$ of $\widetilde{X}$.

We now discuss how $\tau_\varphi(X, \tilde{e})$ depends on the choices made above. Consider the group homomorphism

$$\pi \xrightarrow{\varphi|\pi} \Lambda^* = \mathrm{GL}(1, \Lambda) \hookrightarrow \mathrm{GL}(\Lambda) \to K_1(\Lambda),$$

which we denote (somewhat abusively) by the same letter φ. Recall that $-1 \in K_1(\Lambda)$ is represented by the (1×1)-matrix $[-1]$. Let $\pm\varphi(\pi)$ be the subgroup of $K_1(\Lambda)$ generated by $-1 \in K_1(\Lambda)$ and $\varphi(\pi)$. We claim that the image $\tau_\varphi(X)$ of $\tau_\varphi(X, \tilde{e})$ in $K_1(\Lambda)/\pm\varphi(\pi)$ is independent of the choice of cell orientations and lifts. Indeed, changing the orientation of e_i^k we change the orientation of $\tilde{e}_i^k$, and so $\tau_\varphi(X, \tilde{e})$ is multiplied by -1 (cf. Remark 1.4.1). If we lift e_i^k to $g\tilde{e}_i^k$ rather than to $\tilde{e}_i^k$, where $g \in \pi$, then $\tau_\varphi(X, \tilde{e})$ is multiplied by $\varphi(g)^{(-1)^k}$. Remark 1.4.1 shows that $\tau_\varphi(X) \in K_1(\Lambda)/\pm\varphi(\pi)$ neither depends on the order of cells of X used to define $\tau_\varphi(X, \tilde{e})$. This also follows from the equality $K_1(\mathbb{Z}) = \{\pm 1\}$: Every permutation matrix vanishes in $K_1(\mathbb{Z})/\{\pm 1\} \subset K_1(\Lambda)/\{\pm 1\}$, and so it vanishes in $K_1(\Lambda)/\pm\varphi(\pi)$.

The torsion $\tau_\varphi(X)$ was introduced by Reidemeister [28] for 3-manifolds. As an application he obtained a classification of 3-dimensional lens spaces up to homeomorphism. Franz [14] extended this invariant to higher dimensions and used it to classify lens spaces in all dimensions. The invariant $\tau_\varphi(X)$ is called a *Reidemeister torsion* or a *Reidemeister–Franz torsion* of X. Let us stress that it is defined only under the acyclicity condition stated above. Recall from Lemma 3.3 that if Λ is a field, then $K_1(\Lambda) = \Lambda^*$ so that $\tau_\varphi(X) \in \Lambda^*/\pm\varphi(\pi_1(X))$.

A *subdivision* of a CW-complex X_1 is a CW-complex X_2 such that X_1 and X_2 have the same underlying topological space and such that each open cell of X_2 is contained in a (possibly higher dimensional) open cell of X_1.

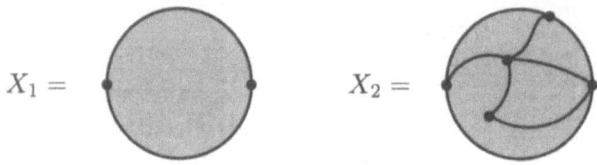

$$X_1 = \qquad\qquad X_2 =$$

Figure 6.1: Subdivision of a CW-complex.

Theorem 6.1 [24] *The torsion $\tau_\varphi(X) \in K_1(\Lambda)/\pm\varphi(\pi_1(X))$ is invariant under cellular subdivision.*

This classical theorem will be sufficient for our purposes. Note, however, a much stronger result due to Chapman [5, 6, 10]: *The torsion is invariant under arbitrary homeomorphisms of CW-complexes.*

Lemma 6.2 (The torsion of S^1) *Let T be a generator of the infinite cyclic group $\pi = \pi_1(S^1)$. Let Λ be a ring as above and let $\varphi: \mathbb{Z}[\pi] \to \Lambda$ be a ring homomorphism. Set $t = \varphi(T) \in \Lambda^*$. Then $H^\varphi_*(S^1) = 0$ if and only if $t - 1 \in \Lambda^*$, in which case $\tau_\varphi(S^1)$ is the image of $(t-1)^{-1}$ in $K_1(\Lambda)/\{\pm t^n\}_{n \in \mathbb{Z}}$.*

In particular, if Λ is a field, then $H^\varphi_*(S^1) = 0$ iff $t \neq 1$, in which case $\tau_\varphi(S^1)$ is the image of $(t-1)^{-1}$ in $\Lambda^*/\{\pm t^n\}_{n \in \mathbb{Z}}$.

Proof. Choose a CW-decomposition of S^1 with two cells e^0, e^1, and orient these cells as in Figure 6.2. Let T correspond to the loop going around S^1 once counterclockwise and let $p: \mathbb{R} \to S^1$ be the universal covering of S^1 given by $x \mapsto e^{2\pi i x}$. Then T acts on the universal cover $\mathbb{R}$ by $T(x) = x + 1$. Choose lifts $\tilde{e}_0$, $\tilde{e}_1$ of e_0, e_1 to $\mathbb{R}$.

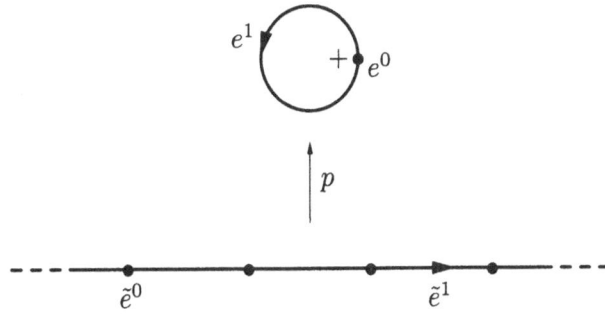

Figure 6.2: The universal covering of S^1.

The only non-vanishing boundary homomorphism in

$$C(\mathbb{R}) = (0 \to \mathbb{Z}[\pi]\tilde{e}^1 \to \mathbb{Z}[\pi]\tilde{e}^0 \to 0)$$

is given by $\partial \tilde{e}^1 = T^{m+1}\tilde{e}^0 - T^m \tilde{e}^0 = T^m(T-1)\tilde{e}^0$ for some $m \in \mathbb{Z}$. (In Figure 6.2, $m = 2$.) The induced boundary map in $C^\varphi(S^1)$ is

$$\partial: \Lambda\tilde{e}^1 \to \Lambda\tilde{e}^0, \quad \tilde{e}^1 \mapsto t^m(t-1)\tilde{e}^0.$$

The chain complex $C^\varphi(S^1)$ is acyclic if and only if $t^m(t-1)$ is invertible in Λ. Since $t^m \in \Lambda^*$, this is equivalent to $t - 1 \in \Lambda^*$. Remark 1.4.3 shows that $\tau_\varphi(S^1) = \tau(C^\varphi(S^1))$ is the image of $(t-1)^{-1}$ in $\Lambda^*/\{\pm t^n\}_{n \in \mathbb{Z}}$. $\square$

6.2 Generalization to pairs

The definition of torsions easily extends to the relative setting of a CW-complex and its CW-subcomplex. Let X be a finite connected CW-complex.

A subset Y of X is called a *subcomplex* if there is a family $\{e_i\}$ of open cells of X such that $Y = \bigcup_i e_i$ and $\bar{e}_i \subset Y$ for all i. Then Y is a closed subset of X and (with the induced topology) Y is a CW-complex with cells $\{e_i\}$. The k-skeleton of Y is $Y^k = Y \cap X^k$. Given a subcomplex Y of X, the pair (X,Y) is called a *CW-pair*. The cellular chain complex $C(Y)$ is clearly a subcomplex of $C(X)$. The quotient $C(X,Y) = C(X)/C(Y)$ is called the *cellular chain complex of the pair* (X,Y). The groups $C_k(X,Y) = C_k(X)/C_k(Y)$ are free abelian groups generated by the oriented open k-cells of X not lying in Y.

Let $p\colon \tilde{X} \to X$ be the universal covering of X. If Y is a subcomplex of X, then $p^{-1}(Y)$ is a subcomplex of $\tilde{X}$. Fixing orientations of the open cells in $X \setminus Y$ we obtain orientations of the open cells in $\tilde{X} \setminus p^{-1}(Y)$. Since the action of $\pi = \pi_1(X)$ on $\tilde{X}$ preserves $p^{-1}(Y)$, we obtain an action of π on $C(\tilde{X}, p^{-1}(Y))$. In this way the cellular chain complex $C(\tilde{X}, p^{-1}(Y))$ becomes a free chain complex over $\mathbb{Z}[\pi]$ with a basis determined by a lift of the (oriented and ordered) open cells of $X \setminus Y$ to $\tilde{X} \setminus p^{-1}(Y)$.

Let $\varphi\colon \mathbb{Z}[\pi] \to \Lambda$ be a ring homomorphism. Set

$$C^\varphi(X,Y) = \Lambda \otimes_\varphi C(\tilde{X}, p^{-1}(Y)).$$

Under the acyclicity condition $H_*^\varphi(X,Y) = H_*(C^\varphi(X,Y)) = 0$ we have a well-defined torsion

$$\tau_\varphi(X,Y) = \tau(C^\varphi(X,Y)) \in K_1(\Lambda)/\pm\varphi(\pi).$$

Theorem 6.1 extends to this case: *The torsion* $\tau_\varphi(X,Y)$ *is invariant under cellular subdivision of the pair* (X,Y).

6.3   The case of commutative rings

Let X be a finite connected CW-complex. Set $\pi = \pi_1(X)$ and $H = H_1(X) = \pi/[\pi,\pi]$. Denote the projection $\mathbb{Z}[\pi] \to \mathbb{Z}[H]$ by h. Let R be a commutative ring with unit $1 \neq 0$. Each ring homomorphism $\varphi\colon \mathbb{Z}[H] \to R$ gives rise to the twisted homology $H_*^{\varphi \circ h}(X)$, and if this homology vanishes, we have the torsion

$$\tau_{\varphi \circ h}(X) \in K_1(R)/\pm(\varphi \circ h)(\pi) = K_1(R)/\pm\varphi(H).$$

In the sequel we shall usually suppress h from the notation and use the symbols $H_*^\varphi(X)$ and $\tau_\varphi(X)$ for $H_*^{\varphi \circ h}(X)$ and $\tau_{\varphi \circ h}(X)$, respectively.

The homology $H_*^\varphi(X)$ and the torsion $\tau_\varphi(X)$ can be computed in terms of the maximal abelian covering $\hat{X}$ of X (see Section 5.4). Proceeding as in Section 6.1, endow $\hat{X}$ with the CW-decomposition induced by X. The action of H on $\hat{X}$ induces an action of $\mathbb{Z}[H]$ on $C_k(\hat{X})$. Lifting ordered oriented cells of X to $\widehat{X}$, we obtain a basis in the free chain complex $R \otimes_\varphi C(\widehat{X})$. The projection $\widetilde{X} \to \widehat{X}$ induces a chain map $C(\widetilde{X}) \to C(\widehat{X})$ and an R-linear chain isomorphism

$$C^{\varphi o h}(X) = R \otimes_{\varphi o h} C(\widetilde{X}) = R \otimes_\varphi C(\widehat{X}).$$

These equalities allow us to compute $H_*^\varphi(X)$ and $\tau_\varphi(X)$ directly from $\widehat{X}$.

Set $G = H/\operatorname{Tors} H$ and denote the projection $\mathbb{Z}[H] \to \mathbb{Z}[G]$ by g. Each ring homomorphism $\psi \colon \mathbb{Z}[G] \to R$ gives rise to the ring homomorphism $\psi \circ g \colon \mathbb{Z}[H] \to R$. If $H_*^{\psi o g}(X)$ vanishes, then the torsion $\tau_{\psi o g}(X)$ is defined. Using the R-linear isomorphisms

$$C^{\psi o g o h}(X) = R \otimes_{\psi o g o h} C(\widetilde{X}) = R \otimes_{\psi o g} C(\widehat{X}) = R \otimes_\psi C(\bar{X})$$

we see that the homology $H_*^{\psi o g}(X)$ and the torsion $\tau_{\psi o g}(X)$ can be computed in terms of the maximal free abelian covering $\bar{X}$ of X.

Similar remarks apply to CW-pairs.

7 The Whitehead torsion

7.1 The Whitehead group

Let π be a group. Consider the composition of group homomorphisms

$$\iota \colon \pi \hookrightarrow (\mathbb{Z}[\pi])^* = \mathrm{GL}(1, \mathbb{Z}[\pi]) \hookrightarrow \mathrm{GL}(\mathbb{Z}[\pi]) \to K_1(\mathbb{Z}[\pi]).$$

The abelian group $K_1(\mathbb{Z}[\pi])/\pm\iota(\pi)$ is called the *Whitehead group of* π and is denoted by $\mathrm{Wh}(\pi)$.

The group $\pm\iota(\pi)$ can be computed as follows. Let $H_1(\pi) = \pi/[\pi,\pi]$ be the first homology group of π and let $q \colon \pi \to H_1(\pi)$ be the projection. Since $K_1(\mathbb{Z}[\pi])$ is abelian, ι factors through q, i.e., there is a group homomorphism $\psi \colon H_1(\pi) \to K_1(\mathbb{Z}[\pi])$ such that $\psi \circ q = \iota$. We claim that ψ is injective. Indeed, the diagram below, in which i is the inclusion, is commutative:

Since i is injective, the claim follows. We may thus identify $\iota(\pi)$ with $H_1(\pi)$. Let

$$\mathrm{aug} \colon \mathbb{Z}[H_1(\pi)] \to \mathbb{Z}, \qquad \sum_{h \in H_1(\pi)} n_h h \mapsto \sum_h n_h$$

be the augmentation. Then $(\mathrm{aug} \circ i \circ q)(\pi) = 1$ and $(\mathrm{aug} \circ \det \circ q_*)([-1]) = -1$. The commutativity of the diagram above implies that $[-1] \notin \iota(\pi)$. Hence

$$\pm\iota(\pi) = (\mathbb{Z}/2\mathbb{Z}) \times H_1(\pi) \subset K_1(\mathbb{Z}[\pi]).$$

Every inner automorphism $\alpha \colon \pi \to \pi$ induces the identity on $\mathrm{Wh}(\pi)$. Indeed, let α be given by $\alpha(a) = bab^{-1}$ for $a, b \in \pi$. The induced automorphism of $\mathrm{GL}(n, \mathbb{Z}[\pi])$ is given by

$$(\lambda_{ij}) \mapsto (b\lambda_{ij}b^{-1}) = (b\,\mathbb{I}_n)(\lambda_{ij})(b\,\mathbb{I}_n)^{-1}$$

where $\mathbb{I}_n$ is the unit $(n \times n)$-matrix. Passing to the abelian group $K_1(\mathbb{Z}[\pi])$, or to its quotient $\mathrm{Wh}(\pi)$, we obtain the identity automorphism.

Since the group operation in $\mathrm{Wh}(\pi)$ is induced by matrix multiplication in $\mathrm{GL}(\mathbb{Z}[\pi])$, we use multiplicative notation for the group operation in $\mathrm{Wh}(\pi)$.

7.2   The Whitehead torsion of deformation retracts

A map $f \colon X \to Y$ between topological spaces is called a *homotopy equivalence* if there exists a (continuous) map $g \colon Y \to X$ such that $g \circ f \colon X \to X$ is homotopic to id_X and $f \circ g \colon Y \to Y$ is homotopic to id_Y. Observe that the composition of homotopy equivalences is again a homotopy equivalence. If X and Y are connected CW-complexes, a map $f \colon X \to Y$ is a homotopy equivalence iff $f_* \colon \pi_i(X) \to \pi_i(Y)$ is an isomorphism for all $i \geq 1$.

Let X be a finite connected CW-complex and let Y be a subcomplex of X which is a deformation retract of X. Then the inclusion $Y \hookrightarrow X$ is a homotopy equivalence. This inclusion induces an isomorphism of $\pi_1(Y, y)$ onto $\pi = \pi_1(X, y)$, where $y \in Y$. Consider the universal covering $p \colon \tilde{X} \to X$. The subcomplex $\tilde{Y} = p^{-1}(Y)$ is a universal covering of Y and a deformation retract of $\tilde{X}$. Therefore, the chain complex $C(\tilde{X}, \tilde{Y})$ over $\mathbb{Z}[\pi]$ is acyclic. Thus, given a ring Λ as in Section 3.2 and a ring homomorphism $\varphi \colon \mathbb{Z}[\pi] \to \Lambda$, we have the torsion $\tau_\varphi(X, Y) = \tau(C^\varphi(X, Y)) \in K_1(\Lambda)/\pm\varphi(\pi)$.

Note that the group ring $\Lambda = \mathbb{Z}[\pi]$ satisfies condition $(*)$ of Section 3.2. Indeed, tensoring with $\mathbb{Z}$ over the augmentation homomorphism $\Lambda \to \mathbb{Z}$ we obtain that an equality $\Lambda^r = \Lambda^s$ implies $r = s$. Therefore, we can consider the torsion of the pair (X, Y) corresponding to the identity homomorphism $id \colon \mathbb{Z}[\pi] \to \mathbb{Z}[\pi]$. Set

$$\tau(X, Y) = \tau_{id}(X, Y) = \tau(C(\tilde{X}, \tilde{Y})) \in K_1(\mathbb{Z}[\pi])/\pm H_1(\pi) = \mathrm{Wh}(\pi).$$

The torsion $\tau(X, Y) \in \mathrm{Wh}(\pi)$ is called the *Whitehead torsion* of the pair (X, Y). As remarked above, $\tau(X, Y)$ is invariant under cellular subdivisions.

Observe that $\tau(X, Y)$ does not depend on the choice of the base point $y \in Y$. To see this, let $y' \in Y$ be another point and let $\gamma_* \colon \mathrm{Wh}(\pi_1(X, y)) \to$

$\mathrm{Wh}(\pi_1(X, y'))$ be the isomorphism induced by the isomorphism $\pi_1(X, y) \to \pi_1(X, y')$ determined by a path γ connecting y and y' in X. It is easy to deduce from definitions that γ_* sends $\tau(X, Y) \in \mathrm{Wh}(\pi_1(X, y))$ to $\tau(X, Y) \in \mathrm{Wh}(\pi_1(X, y'))$. (Note that γ_* does not depend on the choice of γ: If γ' is another path connecting y and y', then the homomorphism

$$\pi_1(X, y) \to \pi_1(X, y), \quad [\alpha] \mapsto [\gamma^{-1} \circ \gamma']^{-1} [\alpha][\gamma^{-1} \circ \gamma']$$

is an inner automorphism, and the discussion in Section 7.1 implies that $\gamma_* = \gamma'_*: \mathrm{Wh}(\pi_1(X, y)) \to \mathrm{Wh}(\pi_1(X, y'))$.)

Let Z be a subcomplex of Y which is a deformation retract of Y. We have inclusions of CW-complexes $Z \subset Y \subset X$ and isomorphisms $\pi_1(Z) = \pi_1(Y) = \pi_1(X) = \pi$. Setting $\widetilde{Y} = p^{-1}(Y)$ and $\widetilde{Z} = p^{-1}(Z)$, we obtain a short exact sequence of acyclic chain complexes over $\mathbb{Z}[\pi]$

$$0 \to C(\widetilde{Y}, \widetilde{Z}) \to C(\widetilde{X}, \widetilde{Z}) \to C(\widetilde{X}, \widetilde{Y}) \to 0.$$

Choosing bases as in Section 6.2 and applying Theorem 3.4, we obtain

$$\tau(X, Z) = \tau(X, Y) \cdot \tau(Y, Z) \in \mathrm{Wh}(\pi). \tag{7.1}$$

Proposition 7.1 (Realization Property) *Let Y be a finite connected CW-complex and let $a \in \mathrm{Wh}(\pi_1(Y))$. Then there is a finite CW-complex X containing Y as a subcomplex such that Y is a deformation retract of X and $\tau(X, Y) = a$.*

Proof. Set $\pi = \pi_1(Y, y)$, where y is a 0-cell of Y. Let (a_{ij}) be an invertible $(k \times k)$-matrix over $\mathbb{Z}[\pi]$ representing a. Fix an integer $n \geq 2$ and consider the wedge $Y' = Y \vee \bigvee_{j=1}^{k} S_j^n$, where S_j^n is an n-dimensional sphere and the wedge is obtained by identifying a point in each S_j^n with y. Of course, $\pi_1(Y', y) = \pi_1(Y, y) = \pi$. Recall that the group $\pi_1(Y', y)$ acts on the n'th homotopy group $\pi_n(Y', y)$ by sweeping n-spheres along loops. In this way, $\pi_n(Y', y)$ becomes a $\mathbb{Z}[\pi]$-module. Denote by $[S_j^n]$ the element of $\pi_n(Y', y)$ represented by S_j^n. For $i = 1, \ldots, k$, set

$$\alpha_i = \sum_{j=1}^{k} a_{ij}[S_j^n] \in \pi_n(Y', y)$$

and realize α_i by a map $f_i : S^n \to Y'$. Let X be the CW-space obtained from Y' by adjoining k copies $D_1^{n+1}, \ldots, D_k^{n+1}$ of an $(n+1)$-dimensional ball along the maps $f_1, \ldots, f_k : S^n \to Y'$, respectively. Set $e_j^n = S_j^n \setminus \{y\} \subset Y' \subset X$ and $e_i^{n+1} = \mathrm{Int}\, D_i^{n+1} \subset X$, $i, j = 1, \ldots, k$. The CW-structure of X is obtained from the one of Y by adding the cells $\{e_j^n\}$ and $\{e_i^{n+1}\}$. The lifts $\{\tilde{e}_j^n\}$ to $\tilde{X}$ form a basis of $C_n(\tilde{X}, \tilde{Y})$ and the lifts $\{\tilde{e}_i^{n+1}\}$ to $\tilde{X}$ form a basis of $C_{n+1}(\tilde{X}, \tilde{Y})$ over $\mathbb{Z}[\pi]$. Hence,

$$
\begin{aligned}
C(\tilde{X}, \tilde{Y}) &= (\ldots \to 0 \to C_{n+1}(\tilde{X}, \tilde{Y}) \to C_n(\tilde{X}, \tilde{Y}) \to 0 \to \ldots) \\
&= (\ldots \to 0 \to (\mathbb{Z}[\pi])^k \to (\mathbb{Z}[\pi])^k \to 0 \to \ldots).
\end{aligned}
$$

If the orientations of the cells e_j^n and e_i^{n+1} and their lifts to $\tilde{X}$ are chosen appropriately, the boundary homomorphism $\partial_n \colon (\mathbb{Z}[\pi])^k \to (\mathbb{Z}[\pi])^k$ is given by the matrix $[a_{ij}]$. Since $[a_{ij}]$ is invertible, $C(\tilde{X}, \tilde{Y})$ is acyclic. Therefore $H_*(\tilde{X}, \tilde{Y}) = 0$. This fact, the equality $\pi_1(X) = \pi_1(Y)$ and elementary homotopy theory imply that Y is a deformation retract of X. The torsion $\tau(X, Y) = \tau(C(\tilde{X}, \tilde{Y}))$ equals $a^{(-1)^{n+1}}$. Taking as n any odd integer ≥ 3, we obtain the claim. $\qquad\square$

7.3   The Whitehead torsion of a homotopy equivalence

Given a continuous map $f \colon X \to Y$ between topological spaces, the *mapping cylinder* M_f of f is built by taking the disjoint union of $X \times [0,1]$ and Y and identifying each point $(x, 1) \in X \times [0,1]$ with $f(x)$:

$$
M_f = (X \times [0,1]) \coprod Y \Big/ (x, 1) = f(x).
$$

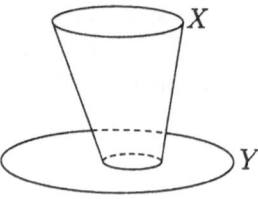

Figure 7.1: The mapping cylinder M_f.

Since the natural projection $(X \times [0,1]) \coprod Y \to M_f$ is an embedding on $X \times 0$ and Y, we shall think of $X = X \times 0$ and Y as subspaces of M_f. As one sees by "sliding along the rays of M_f", the map $p \colon M_f \to Y$, given by

$$
\begin{aligned}
p(x, t) &= f(x) \quad \text{for } x \in X,\, t < 1, \\
p(y) &= y \qquad\quad \text{for } y \in Y
\end{aligned}
$$

is a deformation retraction. This and the commutative diagram

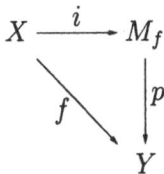

show that the inclusion $i\colon X \hookrightarrow M_f$ is a homotopy equivalence iff f is a homotopy equivalence.

If $f\colon X \to Y$ is a cellular map between CW-complexes, then M_f has a canonical CW-structure which makes X and Y subcomplexes. The open cells of M_f are $\{e\}$, $\{e{\times}]0,1[\}$ and $\{e'\}$, where e runs over open cells in X and e' runs over open cells in Y. (Compare Figure 7.2, where $f(x_1) = y_1$ and f maps the arc $\overline{x_2 x_3}$ to y_2.)

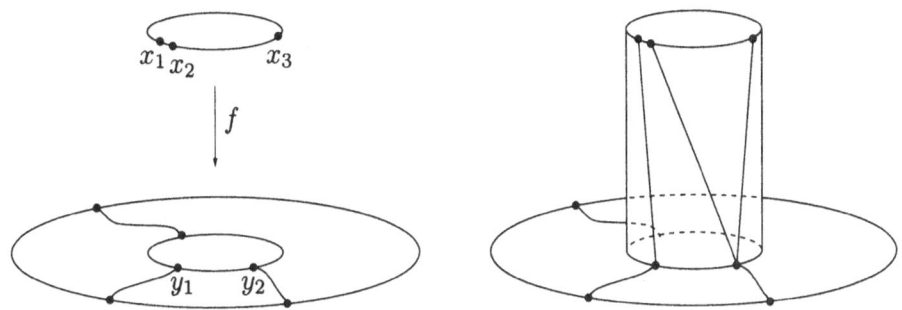

Figure 7.2: The CW-decomposition of M_f.

Now, let X and Y be finite connected CW-spaces and let $f\colon X \to Y$ be a cellular homotopy equivalence. The *Whitehead torsion of f* is defined by

$$\tau(f) = p_*\left(\tau(M_f, X)\right) \in \mathrm{Wh}(\pi_1(Y))$$

where $p_*\colon \mathrm{Wh}(\pi_1(M_f)) \to \mathrm{Wh}(\pi_1(Y)))$ is the isomorphism induced by the isomorphism $p_\#\colon \pi_1(M_f) \to \pi_1(Y)$.

Lemma 7.2 *Let X, Y, Z be finite connected CW-complexes and let $f, f'\colon X \to Y$ and $g\colon Y \to Z$ be cellular homotopy equivalences.*

1. *If f and f' are homotopic, then $\tau(f) = \tau(f')$.*
2. $\tau(g \circ f) = \tau(g) \cdot g_*(\tau(f))$.

A proof may be found in [24, Section 7] or in [6, (22.4)].

Since any continuous map between CW-spaces is homotopic to a cellular map, Lemma 7.2.1 shows that the Whitehead torsion can be defined for arbitrary homotopy equivalences between finite connected CW-complexes.

8 Simple homotopy equivalences

Definition 8.1 A homotopy equivalence $f\colon X \to Y$ of finite connected CW-complexes is said to be *simple* if $\tau(f) = 1 \in \mathrm{Wh}(\pi_1(Y))$.

Lemma 7.2 implies the following theorem.

Theorem 8.2 *A map between finite connected CW-complexes homotopic to a simple homotopy equivalence is itself a simple homotopy equivalence. Compositions of simple homotopy equivalences are simple homotopy equivalences.*

To give a geometric description of simple homotopy equivalences we introduce so-called elementary expansions and elementary collapses.

Let X be a subcomplex of a finite connected CW-complex Y. Assume that Y is obtained from X by adjoining an open k-cell e^k along a map $\partial D^k \to X^{k-1}$ and an open $(k+1)$-cell e^{k+1} along a map $g\colon \partial D^{k+1} \to X^k \cup e^k$ such that $g^{-1}(e^k)$ is an open k-ball in ∂D^{k+1} mapped by g homeomorphically onto e^k. Clearly, X is a deformation retract of Y. In this situation the inclusion $X \hookrightarrow Y$ is called an *elementary expansion*, and its homotopy inverse $Y \to X$ is called an *elementary collapse*. We may think of elementary expansions and collapses as local models for homotopy equivalences.

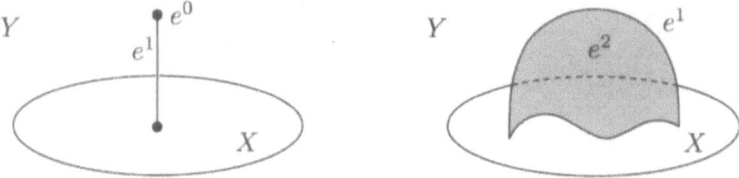

Figure 8.1: Elementary expansions in low dimensions.

Our next aim is to show that elementary expansions and collapses are simple homotopy equivalences. To this end we need a few lemmas.

Lemma 8.3 *Let $X \hookrightarrow Y$ be an elementary expansion. Then $\tau(Y, X) = 1$.*

Proof. Let Y be obtained from X be adjoining two cells e^k and e^{k+1}. If $\tilde{e}^k$ and $\tilde{e}^{k+1}$ are lifts of e^k and e^{k+1} to the universal covering of Y, then

$$C(\tilde{Y}, \tilde{X}) = (\ldots \to 0 \to \mathbb{Z}[\pi]\tilde{e}^{k+1} \xrightarrow{\partial} \mathbb{Z}[\pi]\tilde{e}^k \to 0 \to \ldots),$$

where $\pi = \pi_1(Y)$. The boundary homomorphism ∂ is given by $\partial \tilde{e}^{k+1} = \pm g \tilde{e}^k$ with $g \in \pi$. Hence, $\tau(Y, X) = 1 \in \mathrm{Wh}(\pi)$.   □

Lemma 8.4 *Let $f: X \to Y$ be a cellular map between finite connected CW-complexes. Then $\tau(M_f, Y) = 1$.*

Proof. Set $n = \dim X$. Let $f_k: X^k \to Y$ be the restriction of f to the k-skeleton of X, so that

$$Y = M_{f_{-1}} \subset M_{f_0} \subset M_{f_1} \subset \cdots \subset M_{f_n} = M_f$$

is an increasing sequence of CW-complexes. Each M_{f_k} is obtained from $M_{f_{k-1}}$ by α_k elementary expansions, where α_k is the number of k-cells of X. By (7.1) and Lemma 8.3,

$$\tau(M_f, Y) = \prod_{k=0}^{n} \tau(M_{f_k}, M_{f_{k-1}}) = 1. \qquad \square$$

Taking as f the identity, we obtain

Corollary 8.5 $\tau(X \times [0, 1], X \times 0) = 1$.

Since the CW-pair $(X \times [0, 1], X \times 1)$ is cellularly homeomorphic to $(X \times [0, 1], X \times 0)$, we also have $\tau(X \times [0, 1], X \times 1) = 1$.

The next lemma shows that the Whitehead torsion for deformation retracts is a special case of the Whitehead torsion for homotopy equivalences.

Lemma 8.6 *Let X be a subcomplex of a finite connected CW-complex Y. If the inclusion $f: X \hookrightarrow Y$ is a homotopy equivalence, then $\tau(f) = \tau(Y, X)$.*

Proof. There are inclusions of CW-spaces $X \times 0 \subset X \times [0, 1] \subset M_f$ and $X \times 1 \subset X \times [0, 1] \subset M_f$. By (7.1), Corollary 8.5 and Lemma 8.4,

$$
\begin{aligned}
\tau(f) &= \tau(M_f, X \times 0) \\
&= \tau(M_f, X \times [0, 1]) \cdot \tau(X \times [0, 1], X \times 0) \\
&= \tau(M_f, X \times [0, 1]) \cdot \tau(X \times [0, 1], X \times 1) \\
&= \tau(M_f, X \times 1) \\
&= \tau(M_f, Y) \cdot \tau(Y, X) = \tau(Y, X). \qquad \square
\end{aligned}
$$

Lemmas 8.3 and 8.6 imply that the torsion of an elementary expansion is 1. Corollary 8.5 shows that for any finite connected CW-complex X, $\tau(id_X) = 1$. This and Lemma 7.2.2 imply that given a homotopy equivalence $f: X \to Y$ of finite connected CW-complexes,

$$\tau(f^{-1}) = f_*^{-1}\left(\tau(f)^{-1}\right). \tag{8.1}$$

Since an elementary collapse is the homotopy inverse of an elementary expansion, its torsion is also 1. This constitutes the easy part of the following theorem.

Theorem 8.7 [6, (22.2)] *A homotopy equivalence between finite connected CW-complexes is simple if and only if it can be decomposed into a finite sequence of elementary expansions and elementary collapses.*

We end this subsection with the following useful observation.

Theorem 8.8 *If X' is a CW-subdivision of a finite connected CW-complex X, then the identity maps $X \to X'$ and $X' \to X$ are simple homotopy equivalences.*

Proof. Clearly, the identity map $id\colon X \to X'$ is a cellular map. We provide the interval $[0,1]$ with the CW-structure consisting of two 0-cells $\{0\}$, $\{1\}$ and the 1-cell $]0,1[$. The cylinder $X' \times [0,1]$ with product CW-structure is a subdivision of the mapping cylinder $M_{id\colon X \to X'}$. By the invariance of torsions under cellular subdivisions,

$$\tau(id\colon X \to X') = \tau(M_{id}, X \times 0) = \tau(X' \times [0,1], X' \times 0) = 1.$$

This and Lemma 7.2.2 imply that $\tau(id\colon X' \to X) = 1$.

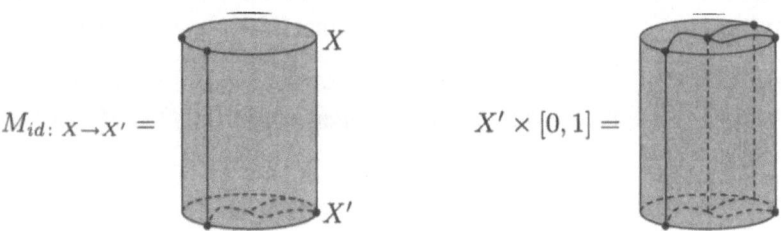

Figure 8.2: Two CW-structures on $X \times [0,1]$.

Again, there is a much stronger result [5, 6, 10] claiming that *homeomorphisms of finite CW-complexes are simple homotopy equivalences.*

Corollary 8.9 *Let $f\colon X \to Y$ be a homotopy equivalence between finite connected CW-complexes. Let X' be a CW-subdivision of X and Y' be a CW-subdivision of Y. Then $\tau(f\colon X \to Y) = \tau(f\colon X' \to Y')$.*

Proof. Present $f\colon X' \to Y'$ as the composition of $id\colon X' \to X$, $f\colon X \to Y$ and $id\colon Y \to Y'$. Theorem 8.8 and Lemma 7.2.2 imply that $\tau(f\colon X' \to Y') = \tau(f\colon X \to Y)$. □

9 Reidemeister torsions and homotopy equivalences

It is easy to check that the Reidemeister torsion is invariant under elementary expansions and elementary collapses of CW-complexes. Hence, Theorem 8.7 implies that the Reidemeister torsion is invariant under simple homotopy equivalences. In this section we establish the following more general result.

Theorem 9.1 *Let* $f\colon X \to Y$ *be a homotopy equivalence of finite connected CW-complexes and let* $\varphi\colon \mathbb{Z}[\pi_1(Y)] \to \Lambda$ *be a ring homomorphism, where* Λ *is a ring as in Section 3.2. Set* $\psi = \varphi \circ f_\# \colon \mathbb{Z}[\pi_1(X)] \to \Lambda$. *Assume that* $H^\varphi_*(Y;\Lambda) = 0$. *Then* $H^\psi_*(X;\Lambda) = 0$ *and*

$$\tau_\varphi(Y) = \tau_\psi(X) \cdot \varphi_*(\tau(f)),$$

where $\varphi_*\colon \mathrm{Wh}(\pi_1(Y)) \to K_1(\Lambda)/\pm\varphi(\pi_1(Y))$ *is the group homomorphism induced by* φ.

Proof. The inclusion $X \subset M_f$ yields a short exact sequence of chain complexes

$$0 \to C(\tilde{X}) \to C(\widetilde{M_f}) \to C(\widetilde{M_f}, \tilde{X}) \to 0.$$

Since this sequence splits, the sequence

$$0 \to \Lambda \otimes_\psi C(\tilde{X}) \to \Lambda \otimes_\varphi C(\widetilde{M_f}) \to \Lambda \otimes_\varphi C(\widetilde{M_f}, \tilde{X}) \to 0$$

is also exact. Since $C(\widetilde{M_f}, \tilde{X})$ is acyclic, $\Lambda \otimes_\varphi C(\widetilde{M_f}, \tilde{X})$ is acyclic. Since X and M_f are homotopy equivalent to Y and, by assumption, $H^\varphi_*(Y;\Lambda) = 0$, the other two complexes are also acyclic. In particular, $H^\psi_*(X;\Lambda) = 0$. The multiplicativity of the torsion implies

$$\tau\big(\Lambda \otimes_\varphi C(\widetilde{M_f})\big) = \tau\big(\Lambda \otimes_\psi C(\tilde{X})\big) \cdot \tau\big(\Lambda \otimes_\varphi C(\widetilde{M_f}, \tilde{X})\big) \qquad (9.1)$$
$$= \tau_\psi(X) \cdot \varphi_*\big(\tau(f)\big).$$

Applying similar arguments to the inclusion $Y \subset M_f$ and using Lemma 8.4 we obtain

$$\tau\big(\Lambda \otimes_\varphi C(\widetilde{M_f})\big) = \tau\big(\Lambda \otimes_\varphi C(\tilde{Y})\big) \cdot \tau\big(\Lambda \otimes_\varphi C(\widetilde{M_f}, \tilde{Y})\big) \qquad (9.2)$$
$$= \tau_\varphi(Y) \cdot \varphi_*\big(\tau(M_f, Y)\big) = \tau_\varphi(Y).$$

Equations (9.1) and (9.2) imply the claim.   □

Corollary 9.2 *If* $f\colon X \to Y$ *is a simple homotopy equivalence, then* $\tau_\psi(X) = \tau_\varphi(Y)$.

10   The torsion of lens spaces

Given an integer $p \geq 3$ and integers $q_1, \ldots, q_n$ relatively prime to p, let $L = L(p; q_1, \ldots, q_n)$ be the lens space defined in Example 5.8.2. We recall that $L = S^{2n-1}/G_p$ where the cyclic group $G_p = \{\zeta \in \mathbb{C} \mid \zeta^p = 1\}$ acts on S^{2n-1} via $\zeta(z_1, \ldots, z_n) = (\zeta^{q_1} z_1, \ldots, \zeta^{q_n} z_n)$.

The lens space L is endowed with the distinguished generator of the fundamental group corresponding to $e^{2\pi i/p} \in G_p$ under the canonical isomorphism $\pi_1(L) = G_p$. By an *esh-equivalence* we mean a simple homotopy equivalence of such "enriched" lens spaces, i.e., a simple homotopy equivalence preserving the distinguished generator of π_1.

In this section, we will study lens spaces up to esh-equivalence. We first observe that some simple transformations of the parameters $q_1, \ldots, q_n$ are induced by esh-equivalences.

(1) Let σ be a permutation of the set $\{1, 2, \ldots, n\}$. Set

$$\varphi_\sigma : \mathbb{C}^n \to \mathbb{C}^n, \quad (z_1, \ldots, z_n) \mapsto (z_{\sigma_1}, \ldots, z_{\sigma_n}).$$

The restriction of φ_σ to S^{2n-1} induces a diffeomorphism

$$L(p; q_1, \ldots, q_n) \to L(p; q_{\sigma_1}, \ldots, q_{\sigma_n}).$$

(2) Since $\zeta^p = 1$ for all $\zeta \in G_p$,

$$L(p; q_1, \ldots, q_n) = L(p; q_1, \ldots, q_i + p, \ldots, q_n).$$

(3) Let $c_i : \mathbb{C}^n \to \mathbb{C}^n$, $(z_1, \ldots, z_n) \mapsto (z_1, \ldots, \overline{z_i}, \ldots, z_n)$ be the complex conjugation of the i'th coordinate. Since $\overline{\zeta^{q_i} z_i} = \zeta^{-q_i} \overline{z_i}$, the restriction of c_i to S^{2n-1} induces a diffeomorphism

$$L(p; q_1, \ldots, q_n) \to L(p; q_1, \ldots, -q_i, \ldots, q_n).$$

Each lens space inherits an orientation from the standard orientation of S^{2n-1}. Note that the diffeomorphisms in (1) and (2) preserve this orientation, while the diffeomorphism in (3) is orientation reversing. However, all these diffeomorphisms preserve the distinguished generator in π_1. We conclude that each lens space is esh-equivalent to a lens space of the form $L(p; q_1, \ldots, q_n)$ with

$$1 \leq q_1 \leq q_2 \leq \cdots \leq q_n < p/2.$$

We call such a lens space *special*.

Theorem 10.1 *Each lens space is esh-equivalent to a unique special lens space. Thus, if there is an esh-equivalence $L(p; q_1, \ldots, q_n) \to L(p; q'_1, \ldots, q'_n)$ between special lens spaces, then $q_i = q'_i$ for $i = 1, \ldots, n$.*

Before proving this theorem we state three useful corollaries.

Corollary 10.2 *Two lens spaces $L(p; q_1, \ldots, q_n)$ and $L(p; q'_1, \ldots, q'_n)$ are esh-equivalent if and only if up to permutation of $q_1, \ldots, q_n$ we have*

$$q'_i = \pm q_i \pmod{p} \quad for \ i = 1, \ldots, n.$$

Corollary 10.3 *Two lens spaces $L(p; q_1, \ldots, q_n)$ and $L(p; q'_1, \ldots, q'_n)$ are simple homotopy equivalent if and only if there is an integer r prime to p such that up to permutation of $q_1, \ldots, q_n$ we have*

$$q'_i = \pm r q_i \pmod{p} \quad for \ i = 1, \ldots, n.$$

Proof. For r prime to p, the identity map $S^{2n-1} \to S^{2n-1}$ induces a diffeomorphism $L(p; rq_1, \ldots, rq_n) \to L(p; q_1, \ldots, q_n)$ sending the distinguished generator to the r'th power of the distinguished generator. Forgetting the distinguished generator is therefore equivalent to considering the numbers $q_1, \ldots, q_n$ up to simultaneous multiplication by integers prime to p. The claim thus follows from Corollary 10.2. □

Each lens space carries a Riemannian metric of constant sectional curvature $+1$ induced by the standard Riemannian metric on S^{2n-1}.

Corollary 10.4 *Simple homotopy equivalent lens spaces are isometric. Esh-equivalent lens spaces are isometric under an isometry preserving the distinguished generator of π_1.*

Proof. The diffeomorphisms considered in (1), (2), (3) and in the proof of Corollary 10.3 are isometries. □

Remark 10.5 By the result cited before Corollary 8.9 or by the results discussed in Section 14, diffeomorphic lens spaces are simple homotopy equivalent. Therefore, the converse of Corollary 10.4 also holds. ◇

The first step in the proof of Theorem 10.1 is the following theorem.

Theorem 10.6 *Let $L = L(p; q_1, \ldots, q_n)$, $p \geq 3$, be a lens space. Let $T \in \pi_1(L)$ be the distinguished generator, and let $r_1, \ldots, r_n \in \mathbb{Z}/p\mathbb{Z}$ be the reciprocal residues defined by $r_i q_i = 1 \pmod{p}$. Let $\mathbb{F}$ be a field and $\varphi: \mathbb{Z}[\pi_1(L)] \to \mathbb{F}$ be a ring homomorphism. Set $t = \varphi(T) \in \mathbb{F}^*$. If $t \neq 1$, then $H_*^\varphi(L) = 0$ and*

$$\tau_\varphi(L) = \prod_{i=1}^{n} (t^{r_i} - 1)^{-1} \in \mathbb{F}^* / \pm \{t^j\}_{j \in \mathbb{Z}/p\mathbb{Z}}.$$

Note that $t^{r_i} \neq 1$ since $(t^{r_i})^{q_i} = t \neq 1$. As it should be, the expression for $\tau_\varphi(L)$ given in Theorem 10.6 is invariant under the transformations of $(q_1, \ldots, q_n)$ discussed in (1), (2) and (3). This is clear for (1) and follows for (2) from $t^p = \varphi(T^p) = \varphi(1) = 1$. If q_i is replaced by $-q_i$, then r_i is replaced by $-r_i$, and we have $t^{-r_i} - 1 = (t^{r_i} - 1) \cdot (-t^{-r_i})$.

Proof of Theorem 10.6. We will first construct a CW-decomposition of L by constructing a G_p-equivariant CW-decomposition of S^{2n-1}. Set $\zeta = e^{2\pi i/p} \in S^1$ and $I_j = [\zeta^j, \zeta^{j+1}] \subset S^1$, where $j \in \mathbb{Z}/p\mathbb{Z}$ (see Figure 10.1).

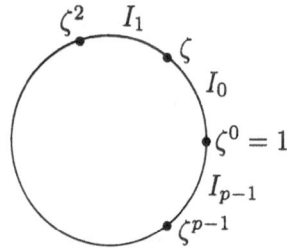

Figure 10.1: The CW-decomposition of S^1.

For $i = 1, \ldots, n$ and $j \in \mathbb{Z}/p\mathbb{Z}$ set

$$
\begin{aligned}
E_j^{2i-2} &= \{(z_1, \ldots, z_n) \in S^{2n-1} \mid z_{i+1} = \cdots = z_n = 0,\ z_i \in \zeta^j \cdot [0,1] \subset \mathbb{C}\} \\
&= \{(z_1, \ldots, z_n) \in S^{2n-1} \mid \sum_{k=1}^{i-1} |z_k|^2 = 1 - |z_i|^2,\ z_i \in \zeta^j \cdot [0,1] \subset \mathbb{C}\}
\end{aligned}
$$

and

$$
\begin{aligned}
E_j^{2i-1} &= \{(z_1, \ldots, z_n) \in S^{2n-1} \mid z_{i+1} = \cdots = z_n = 0,\ z_i \in I_j \cdot [0,1] \subset \mathbb{C}\} \\
&= \{(z_1, \ldots, z_n) \in S^{2n-1} \mid \sum_{k=1}^{i-1} |z_k|^2 = 1 - |z_i|^2,\ z_i \in I_j \cdot [0,1] \subset \mathbb{C}\}.
\end{aligned}
$$

E.g., let $n = 2$. Then

$$
\begin{aligned}
E_j^0 &= (\zeta^j, 0) \in S^3, \\
E_j^1 &= I_j \times 0 \subset S^3, \\
E_j^2 &= \{(z_1, t\zeta^j) \mid 0 \leq t \leq 1,\ |z_1|^2 = 1 - t^2\}, \\
E_j^3 &= \{(z_1, z_2) \in S^3 \mid z_2 \in I_j \cdot [0,1] \subset \mathbb{C}\}.
\end{aligned}
$$

We claim that each E_j^k is a closed k-ball. Indeed, for every $t \in [0,1[$ the equation $\sum_{k=1}^{i-1} |z_k|^2 = 1 - t^2$ determines a $(2i-3)$-sphere. Hence, E_j^{2i-2} is the cone over S^{2i-3}, i.e., a closed $(2i-2)$-ball (compare Figure 10.2).

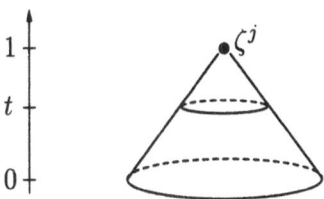

Figure 10.2: The cell E_j^{2i-2}.

As we have just seen, for each $\alpha \in I_j$ the subset of E_j^{2i-1} determined by $z_i \in \alpha \cdot [0,1]$ is a ball. These balls have the same boundary and no other common points. Therefore, E_j^{2i-1} is a closed $(2i-1)$-ball. The structure of E_j^3 is illustrated in Figure 10.3. The origin $\zeta^j \cdot 0 = 0$ corresponds to the common boundary S^1, the intervals $\zeta^j \cdot [0,1]$ and $\zeta^{j+1} \cdot [0,1]$ correspond to D_-^2 and D_+^2, respectively, and if α is the midpoint of I_j, then the interval $\alpha \cdot [0,1]$ corresponds to the equatorial disc E.

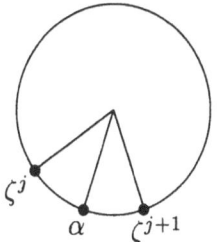 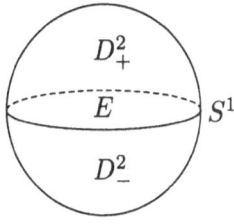

Figure 10.3: The cell E_j^3.

The closed balls

$$\{E_j^{2i-2}, E_j^{2i-1}\}_{\substack{i=1,\ldots,n \\ j \in \mathbb{Z}/p\mathbb{Z}}}$$

form a CW-decomposition of S^{2n-1} with p cells in each dimension. Indeed, the interiors of these balls are all disjoint. Furthermore, the topological boundaries of these balls are given by

$$\begin{aligned}
\partial E_j^{2i-2} &= E_0^{2i-3} \cup E_1^{2i-3} \cup \cdots \cup E_{p-1}^{2i-3}, \\
\partial E_j^{2i-1} &= E_j^{2i-2} \cup E_{j+1}^{2i-2},
\end{aligned} \tag{10.1}$$

where $E_j^{-1} = \emptyset$. The $(2i-1)$-skeleton of this complex equals $S^{2i-1} \subset S^{2n-1}$.

Set $e_j^k = \operatorname{Int} E_j^k$. We orient the open cells e_j^k by induction on k. Let the 0-cells $e_j^0 = (\zeta^j, 0, \ldots, 0)$ be positively oriented. Assume that the cells e_j^k,

$k \leq 2i - 3$, are oriented. We orient e_j^{2i-2} in such a way that

$$\partial e_j^{2i-2} = e_0^{2i-3} + e_1^{2i-3} + \cdots + e_{p-1}^{2i-3}, \tag{10.2}$$

and then orient e_j^{2i-1} so that

$$\partial e_j^{2i-1} = e_{j+1}^{2i-2} - e_j^{2i-2}. \tag{10.3}$$

Observe that with these choices the boundary homomorphisms of the cellular chain complex $C(S^{2n-1})$ are given by (10.2) and (10.3).

Recall that the generator $\zeta = e^{2\pi i/p} \in G_p$ acts on S^{2n-1} by $\zeta \cdot (z_1, \ldots, z_n) = (e^{2\pi i q_1/p} z_1, \ldots, e^{2\pi i q_n/p} z_n)$. An easy induction on the dimension shows that ζ permutes the cells e_j^k in an orientation preserving manner. More precisely,

$$\zeta(e_j^{2i-2}) = e_{j+q_i}^{2i-2} \quad \text{and} \quad \zeta(e_j^{2i-1}) = e_{j+q_i}^{2i-1}. \tag{10.4}$$

By (10.2), (10.3) and (10.4), the boundary homomorphism of the chain complex

$$C(S^{2n-1}) = (\ldots \to \mathbb{Z}[G_p] e_0^k \to \mathbb{Z}[G_p] e_0^{k-1} \to \ldots)$$

is given by

$$\partial e_0^{2i-2} = e_0^{2i-3} + \zeta e_0^{2i-3} + \cdots + \zeta^{p-1} e_0^{2i-3} = \left(\sum_{j=0}^{p-1} \zeta^j \right) e_0^{2i-3},$$

$$\partial e_0^{2i-1} = \zeta^{r_i} e_0^{2i-2} - e_0^{2i-2} = (\zeta^{r_i} - 1) e_0^{2i-2}.$$

Recall that $T \in \pi_1(L)$ corresponds to $\zeta \in G_p$ under the canonical isomorphism $\pi_1(L) = G_p$. Since, by assumption, $t = \varphi(T) \neq 1 \in \mathbb{F}^*$ and since $t^p = \varphi(T^p) = \varphi(1) = 1$, we have that

$$\sum_{j=0}^{p-1} t^j = \frac{t^p - 1}{t - 1} = 0.$$

Hence, the chain complex $C^\varphi(L) = \mathbb{F} \otimes_{\mathbb{Z}[G_p]} C(S^{2n-1})$ is obtained by concatenation of the chain complexes

$$\left(\xrightarrow{0} C_{2i-1} \xrightarrow{t^{r_i} - 1} C_{2i-2} \xrightarrow{0} \right) = \left(\xrightarrow{0} \mathbb{F} e_0^{2i-1} \xrightarrow{t^{r_i} - 1} \mathbb{F} e_0^{2i-2} \xrightarrow{0} \right)$$

where $i = 1, \ldots, n$. Since $t^{r_i} \neq 1$ this chain complex is acyclic. It follows from definitions that

$$\tau_\varphi(L) = \tau(C^\varphi(L)) = \prod_{i=1}^n (t^{r_n} - 1)^{-1} \in \mathbb{F}^* / \pm \{t^j\}_{j \in \mathbb{Z}/p\mathbb{Z}}.$$

This completes the proof of Theorem 10.6.   □

To prove Theorem 10.1 we need the following number-theoretic result. For a proof we refer to [9, Chapter I].

Lemma 10.7 (Franz Independence Lemma) *Set $S = \{j \in \mathbb{Z}/p\mathbb{Z} \mid j \text{ is prime} \text{ to } p\}$. Suppose that $\{a_j\}_{j \in S}$ is a set of integers satisfying*

(1) $\sum_{j \in S} a_j = 0$,

(2) $a_j = a_{-j}$,

(3) $\prod_{j \in S}(\zeta^j - 1)^{a_j} = 1$ *for every p'th root of unity $\zeta \neq 1$.*

Then $a_j = 0$ for all $j \in S$.

Proof of Theorem 10.1. As we observed at the beginning of this section, each lens space is esh-equivalent to a special lens space. It remains to show that if there is an esh-equivalence $L = L(p; q_1, \ldots, q_n) \to L' = L(p; q_1', \ldots, q_n')$ between special lens spaces, then $q_i = q_i'$ for all i.

Fix a p'th root of unity $\zeta \neq 1$ and denote by φ_ζ the ring homomorphism $\mathbb{Z}[\pi_1(L)] \to \mathbb{C}$ mapping the distinguished generator of $\pi_1(L)$ to ζ. By Theorem 10.6, we have

$$\tau_{\varphi_\zeta}(L) = \prod_{i=1}^{n}(\zeta^{r_i} - 1)^{-1} \in \mathbb{C}^*/\pm\{\zeta^j\}_{j \in \mathbb{Z}/p\mathbb{Z}}$$

where $r_1, \ldots, r_n \in \mathbb{Z}/p\mathbb{Z}$ are the residues reciprocal to $q_1, \ldots, q_n$. A similar formula computes $\tau_{\varphi_\zeta'}(L)$ where $\varphi_\zeta' \colon \mathbb{Z}[\pi_1(L')] \to \mathbb{C}$ is the ring homomorphism sending the distinguished generator of $\pi_1(L')$ to ζ. Assume that $f \colon L \to L'$ is an esh-equivalence. Then, by Corollary 9.2, $\tau_{\varphi_\zeta}(L) = \tau_{\varphi_\zeta'}(L')$. We conclude that

$$\prod_{i=1}^{n}(\zeta^{r_i} - 1) = \pm\zeta^d \prod_{i=1}^{n}(\zeta^{r_i'} - 1) \tag{10.5}$$

for some $d \in \mathbb{Z}/p\mathbb{Z}$ (where $r_i' q_i' = 1 \pmod{p}$ for all i). Multiplying both sides of equality (10.5) by their complex conjugates, we obtain

$$\prod_{i=1}^{n}(\zeta^{r_i} - 1)(\zeta^{-r_i} - 1) = \prod_{i=1}^{n}(\zeta^{r_i'} - 1)(\zeta^{-r_i'} - 1). \tag{10.6}$$

Set $S = \{j \in \mathbb{Z}/p\mathbb{Z} \mid j \text{ is prime to } p\}$. For $j \in S$, denote by m_j the number of terms in the sequence $(r_1, -r_1, \ldots, r_n, -r_n)$ equal to j. It is clear that $m_{-j} = m_j$ and $\sum_{j \in S} m_j = 2n$. Similarly, for $j \in S$, denote by m_j' the number of terms in the sequence $(r_1', -r_1', \ldots, r_n', -r_n')$ equal to j. As above, $m_{-j}' = m_j'$

and $\sum_{j \in S} m'_j = 2n$. Set $a_j = m_j - m'_j$. The set $\{a_j\}_{j \in S}$ satisfies conditions (1) and (2) of Lemma 10.7, and condition (3) follows from (10.6):

$$\prod_{j \in S}(\zeta^j - 1)^{a_j} = \prod_{j \in S}(\zeta^j - 1)^{m_j}(\zeta^j - 1)^{-m'_j}$$

$$= \prod_{i=1}^{n}(\zeta^{r_i} - 1)(\zeta^{-r_i} - 1)(\zeta^{r'_i} - 1)^{-1}(\zeta^{-r'_i} - 1)^{-1} = 1.$$

By Lemma 10.7, $a_j = 0$, and so $m_j = m'_j$ for all $j \in S$. Hence, under some reordering $r'_{k_1}, \ldots, r'_{k_n}$ of the sequence $r'_1, \ldots, r'_n$, we have $r'_{k_i} = \pm r_i \pmod{p}$ for $i = 1, \ldots, n$. Then $q'_{k_i} = \pm q_i \pmod{p}$. Since we assumed that

$$1 \le q_1 \le q_2 \le \cdots \le q_n < p/2 \quad \text{and} \quad 1 \le q'_1 \le \cdots \le q'_n < p/2,$$

we conclude that $k_i = i$ and $q'_i = q_i$ for $i = 1, \ldots, n$. This concludes the proof of Theorem 10.1.   □

In the remaining part of this section we compare the esh-classification of lens spaces with their homotopy classification.

Exercise 10.8 Prove that all $(2n - 1)$-dimensional lens spaces with the same fundamental group have the same homology and homotopy groups.   ◇

The next theorem shows that besides the dimension and the order of π_1, the lens spaces have only one additional homotopy invariant. By an *eh-equivalence* between lens spaces we mean a homotopy equivalence preserving the distinguished generator of π_1.

Theorem 10.9 (Homotopy classification, [9, p. 96–101] or [6, (29.6)]) *Two lens spaces $L(p; q_1, \ldots, q_n)$ and $L(p; q'_1, \ldots, q'_n)$ are eh-equivalent if and only if $q'_1 \cdots q'_n = \pm q_1 \cdots q_n \pmod{p}$. Hence, these lens spaces are homotopy equivalent if and only if $q'_1 \cdots q'_n = \pm r^n q_1 \cdots q_n \pmod{p}$ for some integer r prime to p.*

Example 10.10 For $p = 3, 4, 6$ and any $n \ge 2$, all lens spaces $L(p; q_1, \ldots, q_n)$ are esh-equivalent to $L(p; 1, \ldots, 1)$.

Consider in more detail 3-dimensional lens spaces with $p = 5, 7$. For $p = 5$, each lens space $L(5; q_1, q_2)$ is esh-equivalent to either $L(5; 1, 1)$ or $L(5; 1, 2)$ or $L(5; 2, 2)$. Using the theorems above it is easy to check that the equivalence classes of simple homotopy equivalence, eh-equivalence and homotopy equivalence *coincide* and are represented by $L(5; 1, 1)$ and $L(5; 1, 2)$. For $p = 7$, each lens space $L(7; q_1, q_2)$ is esh-equivalent to exactly one of

$$L(7; 1, 1), \quad L(7; 1, 2), \quad L(7; 1, 3), \quad L(7; 2, 2), \quad L(7; 2, 3), \quad L(7; 3, 3).$$

There are two simple homotopy equivalence classes:

$$L(7;1,1) \sim L(7;2,2) \sim L(7;3,3), \quad L(7;1,2) \sim L(7;1,3) \sim L(7;2,3),$$

and there are three eh-equivalence classes:

$$L(7;1,1) \approx L(7;2,3), \quad L(7;1,2) \approx L(7;3,3), \quad L(7;1,3) \approx L(7;2,2).$$

Finally, all these lens spaces are homotopy equivalent to each other.

Solution of Exercise 10.8. Since the CW-decomposition of S^{2n-1} constructed in the proof of Theorem 10.6 is G_p-equivariant, the projection $\pi\colon S^{2n-1} \to L = L(p;q_1,\ldots,q_n)$ induces a CW-decomposition of L with one cell in each dimension. The cells are the sets $e^k = \pi(e_j^k)$, where $j \in \mathbb{Z}/p\mathbb{Z}$ and $k = 0,\ldots,2n-1$. The orientation of e_0^k induces an orientation of e^k. We compute

$$
\begin{aligned}
\partial e^{2i-2} &= \partial \pi(e_0^{2i-2}) = \pi \partial(e_0^{2i-2}) \\
&= \pi(e_0^{2i-3} + \cdots + \zeta^{p-1} e_0^{2i-3}) = p\, e^{2i-3}
\end{aligned}
$$

and

$$\partial e^{2i-1} = \pi \partial(e_0^{2i-1}) = \pi(\zeta^{r_i} e_0^{2i-2} - e_0^{2i-2}) = 0.$$

It follows that

$$C(L) = (0 \to C_{2n-1} \xrightarrow{0} C_{2n-2} \xrightarrow{\cdot p} C_{2n-3} \xrightarrow{0} \cdots \xrightarrow{\cdot p} C_1 \xrightarrow{0} C_0 \to 0),$$

where $C_k = \mathbb{Z}e^k$ for all $k = 0,1,\ldots 2n-1$. Therefore,

$$
H_i(L;\mathbb{Z}) = \begin{cases}
\mathbb{Z}, & i = 0, 2n-1, \\
0, & i \text{ even and } i \neq 0, \\
\mathbb{Z}/p\mathbb{Z}, & i \text{ odd and } 1 \leq i \leq 2n-3.
\end{cases}
$$

Since for any covering $\widetilde{X} \to X$ the induced map $\pi_k(\widetilde{X}) \to \pi_k(X)$ is an isomorphism for $k \geq 2$, the higher homotopy groups of L are equal to those of S^{2n-1}.

11 Milnor's torsion and Alexander's function

11.1 Preliminaries on free abelian groups

Let G be a free abelian group with $b \geq 1$ generators $t_1,\ldots,t_b$. The group ring $\mathbb{Z}[G]$ is isomorphic to the ring of Laurent polynomials on b indeterminates $\mathbb{Z}[t_1^{\pm 1},\ldots,t_b^{\pm 1}]$. Since $\mathbb{Z}$ is a domain, the polynomial ring $\mathbb{Z}[t_1,\ldots,t_b]$ is a domain. Its invertible elements are ± 1. For every $f \in \mathbb{Z}[G]$ there is a monomial

$t_1^{k_1} \cdots t_b^{k_b} \in G$ such that $t_1^{k_1} \cdots t_b^{k_b} \cdot f \in \mathbb{Z}[t_1, \ldots, t_b]$. This implies that $\mathbb{Z}[G]$ is a domain with invertible elements

$$(\mathbb{Z}[G])^* = \pm G = \{\pm t_1^{k_1} \cdots t_b^{k_b} \mid k_1, \ldots, k_b \in \mathbb{Z}\}.$$

The ring $\mathbb{Z}[G]$ is not a principal ideal domain. E.g., the ideal $(2, t_1 - 1)$ is not a principal ideal. However, since $\mathbb{Z}[t_1, \ldots, t_b]$ is a unique factorization domain, so is $\mathbb{Z}[G]$. Furthermore, by the Hilbert Basissatz, $\mathbb{Z}[t_1, \ldots, t_b]$ is Noetherian, and so $\mathbb{Z}[G]$ is Noetherian.

Since $\mathbb{Z}[G]$ is a domain, we can consider its quotient field. It is denoted by $Q(G)$. Clearly $\mathbb{Z}[G] \subset Q(G)$.

11.2 The Alexander function

Let X be a finite connected CW-complex. Set $H = H_1(X; \mathbb{Z})$ and $G = H/\operatorname{Tors} H$. Recall the tower of coverings $\hat{X} \to \bar{X} \to X$ where $\hat{X}$ (resp. $\bar{X}$) is the maximal abelian (resp. maximal free abelian) covering of X with group of covering transformations H (resp. G).

Definition 11.1 The $\mathbb{Z}[G]$-module $A_i(X) = H_i(\bar{X}; \mathbb{Z})$ is called the *i'th Alexander module of* X.

Thus, $A_i(X) = H_i(C)$ where $C = C(\bar{X})$ is a free chain complex of finite rank over $\mathbb{Z}[G]$. Since $\mathbb{Z}[G]$ is Noetherian, the module $A_i(X)$ is finitely generated over $\mathbb{Z}[G]$ for all $i \geq 0$. This allows us to consider the elementary ideals of $A_i(X)$, see Section 4.1. Recall that $\operatorname{ord} A_i(X) \in \mathbb{Z}[G]/\pm G$ is the generator of the smallest principal ideal in $\mathbb{Z}[G]$ containing the elementary ideal $E_0(A_i(X)) \subset \mathbb{Z}[G]$. By Remark 4.5.2,

$$\operatorname{ord} A_i(X) \neq 0 \iff \operatorname{rk}_{\mathbb{Z}[G]} A_i(X) = 0 \iff Q(G) \otimes_{\mathbb{Z}[G]} A_i(X) = 0.$$

Definition 11.2 If $\operatorname{ord} A_i(X) \neq 0$ for all i, then the product

$$A(X) = \prod_{i=0}^{\dim X} (\operatorname{ord} A_i(X))^{(-1)^{i+1}} \in Q(G)/\pm G$$

is called the *Alexander function of* X. If $\operatorname{ord} A_i(X) = 0$ for some i, then we set $A(X) = 0 \in Q(G)/\pm G$.

11.3 Milnor's torsion

We continue to use the notation of the previous subsection. Consider the ring homomorphism

$$\mu \colon \mathbb{Z}[H] \xrightarrow{g} \mathbb{Z}[G] \hookrightarrow Q(G)$$

where $g \colon \mathbb{Z}[H] \to \mathbb{Z}[G]$ is the natural projection.

Definition 11.3 If $H_*^\mu(X) = 0$, then the *Milnor torsion* of X is the torsion

$$\tau_\mu(X) \in K_1(Q(G))/\pm\mu(H) = (Q(G))^*/\pm G \subset Q(G)/\pm G.$$

If $H_*^\mu(X) \neq 0$, we set $\tau_\mu(X) = 0 \in Q(G)/\pm G.$

Observe that

$$
\begin{aligned}
H_i^\mu(X) &= H_i\left(Q(G) \otimes_\mu C(\widehat{X})\right) \\
&= Q(G) \otimes_{\mathbb{Z}[G]} H_i\left(\mathbb{Z}[G] \otimes_g C(\widehat{X})\right) \\
&= Q(G) \otimes_{\mathbb{Z}[G]} H_i\left(C(\bar{X})\right) \\
&= Q(G) \otimes_{\mathbb{Z}[G]} A_i(X).
\end{aligned}
$$

We conclude that $\tau_\mu(X) \neq 0$ if and only if ord $A_i(X) \neq 0$ for all i.

Theorem 11.4 [34] *Let X be a finite connected CW-complex. Then*

$$\tau_\mu(X) = A(X).$$

Proof. If $\tau_\mu(X) = 0$, then ord $A_i(X) = 0$ for some i, and so, by definition, $A(X) = 0$. Assume that $\tau_\mu(X) \neq 0$. Then rk $A_i(X) = 0$ for all i. Since $C(\bar{X})$ is a based free chain complex of finite rank over the Noetherian unique factorization domain $\mathbb{Z}[G]$, we are in the situation of Theorem 4.7, and the claim follows. □

Corollary 11.5 *The Milnor torsion is a homotopy invariant.*

This follows from Theorem 11.4 and the homotopy invariance of the Alexander function.

As an exercise, the reader may deduce the corollary from Theorem 9.1.

Remarks 11.6

1. As we saw in Section 9, the Reidemeister torsions in general are not invariant under arbitrary homotopy equivalences. The Milnor torsion, which is a special instant of a Reidemeister torsion, is a rougher invariant in this respect.

2. The Alexander function and the Milnor torsion of X are interesting only if $b_1(X) = \dim H_1(X; \mathbb{R}) \geq 1$. If $b_1(X) = 0$, then ord $A_0(X) = 0$ and therefore $A(X) = \tau_\mu(X) = 0$.

11.4 Computation of the Milnor torsion

Theorem 11.4 allows to compute the Milnor torsion using homological methods. In this subsection we collect several results in this direction.

Consider a finite connected n-dimensional CW-complex X with $b_1(X) \geq 1$
such that $\operatorname{ord} A_i(X) \neq 0$ for all i. By definition $A(X) = \prod_{i=0}^{n}(\operatorname{ord} A_i(X))^{(-1)^{i+1}}$.
The highest and lowest terms of this product can be computed as follows.

By Remark 4.5.2, the assumption $\operatorname{ord} A_i(X) \neq 0$ implies that

$$A_i(X) = \operatorname{Tors}_{\mathbb{Z}[G]} A_i(X).$$

Clearly, $A_n(X) = \operatorname{Ker}(\partial\colon C_n(\bar{X}) \to C_{n-1}(\bar{X}))$, and so $A_n(X)$ is a submodule
of the free $\mathbb{Z}[G]$-module $C_n(\bar{X})$. Thus, $A_n(X) = \operatorname{Tors}_{\mathbb{Z}[G]} A_n(X) = 0$ and so

$$\operatorname{ord} A_n(X) = 1. \tag{11.1}$$

We next compute $\operatorname{ord} A_0(X)$. Since X is connected, so is $\bar{X}$, and hence
$A_0(X) = H_0(\bar{X}; \mathbb{Z}) = \mathbb{Z}$. Observe that G acts trivially on $A_0(X)$. Indeed, let
$x \in \bar{X}$ be a point and let $g \in G$. The points x and gx can be joined by a path
in $\bar{X}$, and so they are homologous. The module $A_0(X)$ is thus the $\mathbb{Z}[G]$-module
on one generator $[x]$ subject to the relations $\{g[x] = [x] \,|\, g \in G\}$. The matrix
of the resulting presentation

$$(\mathbb{Z}[G])^{|G|} \to \mathbb{Z}[G] \to \mathbb{Z} = A_0(X) \to 0$$

of $A_0(X)$ is thus an infinite column vector with entries $g - 1$, $g \in G$. The
elementary ideal $E_0(\mathbb{Z})$ is therefore the ideal of $\mathbb{Z}[G]$ generated by $\{g - 1 \,|\, g \in G\}$. Choose generators $t_1, \ldots, t_b$ of G, where $b = b_1(X) \geq 1$. If $b = 1$, we set
$t = t_1$. We then find that

$$\operatorname{ord} A_0(X) \quad = \quad \gcd \langle t_1^{m_1} \cdots t_b^{m_b} - 1 \rangle_{m_1, \ldots, m_b \in \mathbb{Z}} \tag{11.2}$$
$$= \quad \begin{cases} t - 1 & \text{if} \quad b_1(X) = 1, \\ 1 & \text{if} \quad b_1(X) \geq 2. \end{cases}$$

For $b_1(X) = 1$, the equality $t^{-1} - 1 = -t^{-1}(t - 1)$ confirms that $\operatorname{ord} A_0(X) \in \mathbb{Z}[G]/ \pm G$ does not depend on the choice of t.

Set $\pi = \pi_1(X)$ and $\bar{\pi} = \pi_1(\bar{X})$. Clearly, $\bar{\pi}$ is the kernel of the natural
projection $\pi \to G$. The $\mathbb{Z}[G]$-module $A_1(X) = H_1(\bar{X}) = \bar{\pi}/[\bar{\pi}, \bar{\pi}]$ thus depends
only on π and the same holds for $\operatorname{ord} A_1(X)$.

Definition 11.7 The order $\operatorname{ord} A_1(X) \in \mathbb{Z}[G]/ \pm G$ is called the *Alexander
polynomial* of $\pi = \pi_1(X)$. It is denoted by Δ_π.

Theorem 11.8 [23] *Let X be a finite connected 2-dimensional CW-complex with $\chi(X) = 0$. If $b_1(X) = 1$, let t be a generator of the infinite cyclic group $G = H_1(X)/\operatorname{Tors} H_1(X)$. Then*

$$\tau_\mu(X) = A(X) = \begin{cases} \Delta_\pi(t-1)^{-1} & \text{if } b_1(X) = 1, \\ \Delta_\pi & \text{if } b_1(X) \geq 2. \end{cases}$$

Proof. Note that $b_1(X) = b_0(X) + b_2(X) - \chi(X) \geq b_0(X) = 1$. If $\Delta_\pi = \operatorname{ord} A_1(X) = 0$, then, by definition, $A(X) = 0$, and the claim follows. Let now $\Delta_\pi \neq 0$. Then $\operatorname{rk} A_1(X) = 0$. By (11.2), $\operatorname{ord} A_0(X) \neq 0$, and so $\operatorname{rk} A_0(X) = 0$. Observe next that

$$\begin{aligned} 0 = \chi(X) &= \sum_i (-1)^i \operatorname{rk}_{\mathbb{Z}[G]} C_i(\bar{X}) \\ &= \sum_i (-1)^i \operatorname{rk}_{\mathbb{Z}[G]} A_i(X) \\ &= \operatorname{rk} A_0(X) - \operatorname{rk} A_1(X) + \operatorname{rk} A_2(X) = \operatorname{rk} A_2(X). \end{aligned}$$

We conclude that $\operatorname{ord} A_2(X) \neq 0$. By Theorem 11.4,

$$\tau_\mu(X) = A(X) = \frac{\operatorname{ord} A_1(X)}{\operatorname{ord} A_0(X) \cdot \operatorname{ord} A_2(X)}.$$

The theorem now follows from (11.1) and (11.2). $\square$

Corollary 11.9 [23] *Let M be a compact connected triangulated 3-manifold whose boundary is non-empty and consists of tori. If $b_1(M) = 1$, let t be a generator of the infinite cyclic group $H_1(M)/\operatorname{Tors} H_1(M)$. Then*

$$\tau_\mu(M) = A(M) = \begin{cases} \Delta_{\pi_1(M)}(t-1)^{-1} & \text{if } b_1(M) = 1, \\ \Delta_{\pi_1(M)} & \text{if } b_1(M) \geq 2. \end{cases}$$

Proof. Let $\mathcal{T}$ be the given triangulation of M. The restriction of $\mathcal{T}$ to ∂M defines a triangulation $\partial \mathcal{T}$ of ∂M. Let

$$\check{M} = \left(M \coprod M\right)/\operatorname{id}_{\partial M}$$

be the double of M (see Figure 11.1). The triangulations $\mathcal{T}$ on both copies of M combine to a triangulation $\check{\mathcal{T}}$ of $\check{M}$. Counting the simplices, edges and vertices in $\mathcal{T}$, $\partial \mathcal{T}$ and $\check{\mathcal{T}}$, we find

$$2\chi(M) = \chi(\check{M}) + \chi(\partial M).$$

Moreover, $\check{M}$ is a closed 3-manifold, and so, by the Poincaré duality with $\mathbb{Z}_2$-coefficients (cf. Section 14 below),

$$\chi(\check{M}) = \sum_{i=0}^{3} (-1)^i \dim_{\mathbb{Z}_2} H_i(\check{M}; \mathbb{Z}_2) = 0.$$

We conclude that $\chi(M) = \frac{1}{2}\chi(\partial M) = 0$.

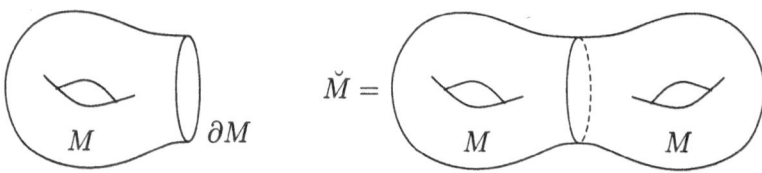

Figure 11.1: M and its double $\check{M}$.

Observe next that M can be deformed onto a 2-dimensional subcomplex. Indeed, take a 2-simplex F in ∂T. There is a unique 3-simplex $T \in \mathcal{T}$ containing F as a face. Clearly, $M^{(1)} = M \setminus (\operatorname{Int} T \cup \operatorname{Int} F)$ is an elementary collapse of M. We can similarly collapse $M^{(1)}$ further until there remain no 3-simplices. In this way we arrive at a 2-dimensional subcomplex $X \subset M$. Since X is a deformation retract of M, we have $\pi_1(X) = \pi_1(M)$ and $\chi(X) = \chi(M) = 0$. Hence, X fulfills the assumptions of Theorem 11.8. The corollary now follows from the homotopy invariance of the Alexander function.   □

The following theorem (together with Corollary 11.9) computes the relative Milnor torsion for 3-manifolds with boundary consisting of tori.

Theorem 11.10 *Under the conditions of Corollary 11.9 we have*

$$\tau_\mu(M, \partial M) = \tau_\mu(M) = A(M).$$

The proof of this theorem is based on the following two lemmas.

Lemma 11.11 *Let T^2 be a 2-torus. Set $H = H_1(T^2)$. Let $\mathbb{F}$ be a field and $\varphi \colon \mathbb{Z}[H] \to \mathbb{F}$ be a ring homomorphism. If $\varphi(H) \neq 1$, then $H_*^\varphi(T^2) = 0$ and $\tau_\varphi(T^2) = 1 \in \mathbb{F}^* / \pm \varphi(H)$.*

Proof. Choose a CW-decomposition of T^2 consisting of one 0-cell e^0, two 1-cells e_1^1 and e_2^1 and one 2-cell e^2 oriented as in Figure 11.2. Let h_1 and h_2 be the generators of H represented by e_1^1 and e_2^1, respectively. With a suitable choice of natural bases $c_1^0, (c_1^1, c_2^1), c^2$ the boundary homomorphisms ∂_0 and ∂_1 in

$$C^\varphi(T^2) = (0 \to C_2 \xrightarrow{\partial_1} C_1 \xrightarrow{\partial_0} C_0 \to 0)$$

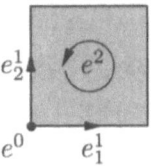

Figure 11.2

are given by $\begin{pmatrix} \varphi(h_1) - 1 \\ \varphi(h_2) - 1 \end{pmatrix}$ and $(1 - \varphi(h_2),\ \varphi(h_1) - 1)$, respectively. By assumption, $\varphi(h_1) \neq 1$ or $\varphi(h_2) \neq 1$ and so $C^\varphi(T^2)$ is acyclic. Assume for concreteness that $\varphi(h_1) \neq 1$. Then $\partial_1 c^2$ generates $\operatorname{Im} \partial_1$ and $\partial_0 c_1^1$ generates $\operatorname{Im} \partial_0 = C_0$. Thus,

$$
\begin{aligned}
\tau(C^\varphi(T^2)) &= \left[\partial_0 c_1^1/c^0\right]^{-1} \left[\partial_1 c^2, c_1^1/c_1^1, c_2^1\right] \left[c^2/c^2\right]^{-1} \\
&= (\varphi(h_1) - 1)^{-1} \begin{vmatrix} 1 - \varphi(h_2) & \varphi(h_1) - 1 \\ 1 & 0 \end{vmatrix} = -1. \qquad \square
\end{aligned}
$$

Lemma 11.12 *Let M be a compact orientable 3-manifold whose boundary contains a torus T^2. Then the inclusion homomorphism*

$$H_1(T^2) \to H_1(M)/\operatorname{Tors} H_1(M)$$

does not vanish.

Proof. Let $a, b \subset T^2$ be two simple closed curves on T^2 transversally intersecting in one point. Assume that their homology classes $[a], [b] \in H_1(M)$ both lie in $\operatorname{Tors} H_1(M)$. Then there are integers $m, n \geq 1$ such that $m[a] = n[b] = 0$. Choose a 2-chain $D \in C_2(M; \mathbb{Z})$ with $\partial D = nb$. This chain represents a relative homology class $[D] \in H_2(M, \partial M)$. The intersection number of $m[a] \in H_1(M)$ with $[D]$ equals $mn \neq 0$. This contradicts $m[a] = 0$. $\qquad \square$

Proof of Theorem 11.10. Let $T_1, \ldots, T_l$ be the components of ∂M. Let μ_i be the composition of the inclusion homomorphism $\mathbb{Z}[H_1(T_i)] \to \mathbb{Z}[H_1(M)]$ with $\mu \colon \mathbb{Z}[H_1(M)] \to Q(H_1(M)/\operatorname{Tors} H_1(M))$. Lemmas 11.11 and 11.12 imply that $H_*^{\mu_i}(T_i) = 0$ and $\tau_{\mu_i}(T_i) = 1$ for all i. Therefore the natural projection $C^\mu(M) \to C^\mu(M, \partial M)$ induces an isomorphism of homologies. If both complexes have non-trivial homology, then $\tau_\mu(M, \partial M) = \tau_\mu(M) = 0$. If both complexes are acyclic, then, by Theorem 1.5,

$$\tau_\mu(M) = \tau_\mu(M, \partial M) \prod_{i=1}^{l} \tau_{\mu_i}(T_i) = \tau_\mu(M, \partial M). \qquad \square$$

12   Group rings of finitely generated abelian groups

12.1   Preliminaries on commutative rings

Let $\{R_i\}_{i=1}^n$ be a finite family of rings. The *direct sum* $R = \oplus_{i=1}^n R_i$ is the set $\times_{i=1}^n R_i$ endowed with coordinate-wise addition and multiplication. Each ring R_i canonically embeds in R and is an ideal in R.

Lemma 12.1 *Let R be a ring which splits as a direct sum of finitely many domains. Then this splitting of R is unique.*

Proof. Let $R = \oplus_{i=1}^m R_i = \oplus_{j=1}^n R_j'$ be two splittings of R, where R_i and R_j' are domains. We need to show that each R_j' is equal to some R_i. Let $a \in R_j'$. Then $a = r_1 + r_2 + \cdots + r_m$, where $r_i \in R_i$. Assume that there are $i \neq i'$ such that $r_i \neq 0$ and $r_{i'} \neq 0$. Let 1_{R_i} and $1_{R_{i'}}$ be the units in R_i and $R_{i'}$, respectively. We have

$$1_{R_i} \cdot r_{i'} = \left\{ \begin{array}{ll} r_{i'} & \text{if} \quad i = i', \\ 0 & \text{if} \quad i \neq i'. \end{array} \right.$$

Since R_j' is an ideal, we have that $r_i = 1_{R_i} \cdot a \in R_j' \setminus \{0\}$ and $r_{i'} = 1_{R_{i'}} \cdot a \in R_j' \setminus \{0\}$. But R_j' is a domain, and so $r_i r_{i'} \neq 0$, a contradiction. This shows that every $a \in R_j'$ lies in some R_i. Let $a, b \in R_j' \setminus \{0\}$. Then there are i and i' such that $a \in R_i$ and $b \in R_{i'}$. Since R_j' is a domain, $ab \neq 0$, and so $i = i'$. We conclude that $R_j' \subset R_i$. By symmetry, $R_j' = R_i$. The lemma now follows. $\square$

Let now R be a commutative ring with $1 \neq 0$. The *classical ring of quotients* $Q(R)$ is the localization of R in the set of its non-zero-divisors. In other words, $Q(R)$ is obtained from R by inverting all non-zero-divisors. The ring $Q(R)$ is a commutative ring with unit. It follows from definitions that the natural ring homomorphism $R \to Q(R)$ is injective. In the sequel we consider R as a subring of $Q(R)$. If R is a domain, then $Q(R)$ is the quotient field of R, and if $\mathbb{F}$ is a field, then $Q(\mathbb{F}) = \mathbb{F}$.

Lemma 12.2 *Let R_1 and R_2 be commutative rings and $R = R_1 \oplus R_2$. Then $Q(R) = Q(R_1) \oplus Q(R_2)$.*

Proof. Given $a \in R$, write $a = a_1 + a_2$ with $a_1 \in R_1$, $a_2 \in R_2$. Clearly, a is a non-zero-divisor in R if and only if both a_1 and a_2 are non-zero-divisors in R_1 and R_2, respectively. The claim thus follows. $\square$

Corollary 12.3

   1. *If $R = \oplus_{i=1}^k R_i$ is a direct sum of domains, then $Q(R) = \oplus_{i=1}^k Q(R_i)$ is a direct sum of fields.*

   2. *If R is a direct sum of fields, then $Q(R) = R$.*

Let $\psi \colon R \to \mathbb{F}$ be a ring homomorphism from a commutative ring with unit R to a field $\mathbb{F}$. Let N_ψ be the set of non-zero-divisors $y \in R$ with $\psi(y) \neq 0$. Clearly, N_ψ is closed under multiplication. Let $Q_\psi(R)$ be the localization of R in N_ψ, i.e.,

$$Q_\psi(R) = \{x \in Q(R) \,|\, \text{there exists } y \in N_\psi \text{ such that } xy \in R\}\,.$$

Then $Q_\psi(R)$ is a subring of $Q(R)$ containing R:

$$R \subset Q_\psi(R) \subset Q(R).$$

Lemma 12.4 *The ring homomorphism $\psi \colon R \to \mathbb{F}$ extends uniquely to a ring homomorphism $\tilde{\psi} \colon Q_\psi(R) \to \mathbb{F}$.*

Proof. Given $x \in Q_\psi(R)$, pick $y \in N_\psi$ such that $xy \in R$. If $\tilde{\psi}$ exists, we must have $\tilde{\psi}(x)\,\psi(y) = \tilde{\psi}(xy) = \psi(xy)$, and so, using $\psi(y) \neq 0$,

$$\tilde{\psi}(x) = \psi(xy)\,\psi(y)^{-1}.$$

The extension is thus unique if it exists. To prove existence we define $\tilde{\psi}(x)$ by the latter formula. The definition of $\tilde{\psi}(x)$ does not depend on the choice of y. Indeed, let y' be another element in N_ψ with $xy' \in R$. Then $\psi(xy)\,\psi(y') = \psi(xyy') = \psi(xy')\,\psi(y)$, and so $\psi(xy)\,\psi(y)^{-1} = \psi(xy')\,\psi(y')^{-1}$. It is straightforward to check that $\tilde{\psi}$ is a ring homomorphism.   □

12.2 Group rings of abelian groups

Let H be a group. For a ring R, set

$$R[H] = \left\{ \sum_{h \in H} q_h h \,\Big|\, q_h \in R,\ q_h = 0 \text{ for all but finitely many } h \right\}.$$

With the addition

$$\left(\sum q_h h \right) + \left(\sum q_h' h \right) = \sum (q_h + q_h') h$$

and the multiplication

$$\left(\sum q_g g \right) \cdot \left(\sum q_h h \right) = \sum (q_g q_h)(gh)$$

the set $R[H]$ becomes an associative ring.

Assume from now on that H is an abelian group. Then the ring $\mathbb{Q}[H]$ is commutative and we can consider its classical ring of quotients $Q(\mathbb{Q}[H])$. We shall denote it by $Q(H)$. Observe that

$$Q(H) = Q(\mathbb{Q}[H]) = Q(\mathbb{Z}[H]) \supset \mathbb{Q}[H] \supset \mathbb{Z}[H].$$

The notation $Q(H)$ is consistent with the one introduced in Section 11.1: If H is torsion-free, then $Q(H)$ is the quotient field of the domains $\mathbb{Z}[H]$ and $\mathbb{Q}[H]$. The goal of this section is to show that for any finitely generated abelian group H, the ring $Q(H)$ splits as a direct sum of fields.

We first analyze the structure of $\mathbb{Q}[H]$ for a finite abelian group H. Recall that a *cyclotomic field* $\mathbb{F}$ is a field extension of $\mathbb{Q}$ by a root of unity, i.e., $\mathbb{F} = \mathbb{Q}\left(e^{2\pi i/m}\right) \subset \mathbb{C}$ for some $m \in \mathbb{N}$.

Theorem 12.5 *Let H be a finite abelian group. Then $\mathbb{Q}[H]$ splits in a unique way as a direct sum of finitely many cyclotomic fields.*

Proof. The uniqueness follows from Lemma 12.1.

Set $\mathbb{F}_m = \mathbb{Q}\left(e^{2\pi i/m}\right)$. The field $\mathbb{Q}$ canonically embeds in $\mathbb{F}_m$, and $\mathbb{F}_m$ is a vector space over $\mathbb{Q}$ with basis consisting of the primitive m'th roots of unity. Recall that the *Galois group* $\mathrm{Aut}_{\mathbb{Q}}(\mathbb{F}_m)$ is the group of field automorphisms of $\mathbb{F}_m$ extending the identity on $\mathbb{Q}$. Each such automorphism maps $e^{2\pi i/m}$ to a primitive m'th root of unity, and conversely each primitive m'th root of unity determines thus an element in $\mathrm{Aut}_{\mathbb{Q}}(\mathbb{F}_m)$. We conclude that

$$|\mathrm{Aut}_{\mathbb{Q}}(\mathbb{F}_m)| \;=\; \dim_{\mathbb{Q}} \mathbb{F}_m \;=\; \#\{n \in \mathbb{Z}/m\mathbb{Z} \,|\, (m,n) = 1\} \qquad (12.1)$$

where $|\mathrm{Aut}_{\mathbb{Q}}(\mathbb{F}_m)|$ denotes the number of elements of $\mathrm{Aut}_{\mathbb{Q}}(\mathbb{F}_m)$.

Let H^* be the set of group homomorphisms $H \to \mathbb{C}^* = \mathbb{C} \setminus \{0\}$. The elements of H^* are called *characters*. Let $g \in H^*$ and $h \in H$. Since H is finite, $g(h) \in \mathbb{C}^*$ is of finite order, and so $g(h) \in S^1$. Hence,

$$H^* = \mathrm{Hom}(H, \mathbb{C}^*) = \mathrm{Hom}(H, S^1).$$

Observe next that

$$|H^*| = |H|. \qquad (12.2)$$

Indeed, if H is a cyclic group of order n, then so is H^*, and if H_1 and H_2 are finite abelian groups, then

$$(H_1 \oplus H_2)^* = H_1^* \oplus H_2^*.$$

Equation (12.2) thus follows from the fact that every finite abelian group is a direct sum of cyclic groups.

Every character $\sigma \in H^*$ extends to a ring homomorphism $\tilde{\sigma} \colon \mathbb{Q}[H] \to \mathbb{C}$ by

$$\sum_{h \in H} q_h h \,\mapsto\, \sum_{h \in H} q_h \sigma(h).$$

Since H is finite, $\sigma(H)$ is a finite subgroup of S^1, i.e., a finite cyclic group. Hence, there is a non-negative integer m_σ such that

$$\sigma(H) = \{\zeta \in \mathbb{C} \,|\, \zeta^{m_\sigma} = 1\}.$$

Then $\tilde{\sigma}(\mathbb{Q}[H]) = \mathbb{F}_{m_\sigma}$.

We say that characters $\sigma, \rho \in H^*$ are *equivalent* and write $\rho \sim \sigma$ if $m_\sigma = m_\rho$ and $\tilde{\sigma}$ differs from $\tilde{\rho}$ by a Galois automorphism of $\mathbb{F}_{m_\sigma}$, i.e., there is $\gamma \in \mathrm{Aut}_{\mathbb{Q}}(\mathbb{F}_{m_\sigma})$ such that the following diagram commutes:

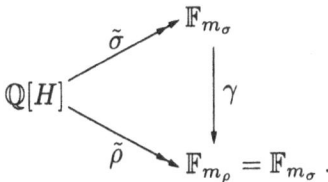

The relation $\sim$ is an equivalence relation on H^*.

The following lemma concludes the proof of Theorem 12.5.

Lemma 12.6 *Let* $\sigma_1, \ldots, \sigma_k \in H^*$ *be representatives of the equivalence classes. Then*

$$\bigoplus_{i=1}^{k} \tilde{\sigma}_i \colon \mathbb{Q}[H] \to \bigoplus_{i=1}^{k} \mathbb{F}_{m_{\sigma_i}}$$

is a ring isomorphism.

Proof. The map $\oplus_{i=1}^{k} \tilde{\sigma}_i$ is clearly a ring homomorphism.

By (12.1), $\dim_{\mathbb{Q}} \mathbb{F}_{m_{\sigma_i}} = |\mathrm{Aut}_{\mathbb{Q}}(\mathbb{F}_{m_{\sigma_i}})|$ is the number of characters equivalent to σ_i. Hence

$$\dim_{\mathbb{Q}} \left(\bigoplus_{i=1}^{k} \mathbb{F}_{m_{\sigma_i}} \right) = \sum_{i=1}^{k} \dim_{\mathbb{Q}} \mathbb{F}_{m_{\sigma_i}}$$

$$= \sum_{i=1}^{k} \#\{\rho \in H^* \,|\, \rho \sim \sigma_i\}$$

$$= |H^*| = |H| = \dim_{\mathbb{Q}} \mathbb{Q}[H].$$

We are left to show that $\oplus_{i=1}^{k} \tilde{\sigma}_i$ is injective. It suffices to prove the following claim: For any $a \in \mathbb{Q}[H]$, $a \neq 0$, there exists $\sigma \in H^*$ such that $\tilde{\sigma}(a) \neq 0$.

Short proof of injectivity. Elementary representation theory tells us that for any finite group Γ the complex group algebra $\mathbb{C}[\Gamma]$ is semisimple and is hence a direct sum of matrix rings. If $\Gamma = H$ is abelian, so is $\mathbb{C}[\Gamma] = \mathbb{C}[H]$, and hence all these matrix rings are 1-dimensional. Thus $\mathbb{C}[H] = \oplus_i \mathbb{C}$. Let $p_i \colon \mathbb{C}[H] \to \mathbb{C}$ be the projection onto the i-summand. Then p_i is a homomorphism of $\mathbb{C}$-algebras. It is therefore the ring extension of a character $H \to \mathbb{C}^*$. We conclude that given a non-zero element $a \in \mathbb{Q}[H] \subset \mathbb{C}[H] = \oplus_i \mathbb{C}$, there is $\sigma \in H^*$ such that $\tilde{\sigma}(a) \neq 0$. $\qquad\square$

Longer but elementary proof of injectivity. We shall give a direct proof of a similar injectivity result, where the coefficient ring $\mathbb{Q}$ of $\mathbb{Q}[H]$ is replaced by $\mathbb{C}$. This will imply our claim.

Suppose first that H is a cyclic group of order n, i.e., $H = \{t^i\}_{i=0}^{n-1}$ and $t^n = 1$. Let $a = \sum_{i=0}^{n-1} a_i t^i \in \mathbb{C}[H]$. Assume that $\tilde{\sigma}(a) = 0$ for all $\sigma \in H^*$. Given an n'th root of unity $\zeta \in \mathbb{C}$, let σ_ζ be the element in H^* with $\sigma_\zeta(t) = \zeta$. Then $a(\zeta) = \tilde{\sigma}_\zeta(a) = 0$. This means that the polynomial $a(t)$ of degree $n-1$ has n distinct complex roots. Hence $a = 0$.

Let now $H = H_1 \oplus H_2$, where H_1 and H_2 are finite abelian groups. Assume by induction that our claim holds for H_1 and H_2. Let $a \in \mathbb{C}[H]$, $a \neq 0$. Then

$$a = \sum_{g \in H_1} \alpha_g \cdot g \quad \text{for certain } \alpha_g \in \mathbb{C}[H_2],$$

and there exists $g_0 \in H_1$ such that $\alpha_{g_0} \neq 0$. By the induction assumption there is $\sigma \in H_2^*$ such that $\tilde{\sigma}(\alpha_{g_0}) \neq 0$. The element σ gives rise to a $\mathbb{C}$-linear homomorphism

$$p_\sigma \colon \mathbb{C}[H] \to \mathbb{C}[H_1], \qquad \sum_{g \in H_1} \beta_g g \mapsto \sum_{g \in H_1} \tilde{\sigma}(\beta_g) g.$$

Since $p_\sigma(a) = \sum_{g \in H_1} \tilde{\sigma}(\alpha_g) g \neq 0$, there exists, by the induction assumption, $\rho \in H_1^*$ such that $\tilde{\rho}\,(p_\sigma(a)) \neq 0$. The element $(\rho, \sigma) \in H_1^* \oplus H_2^* = (H_1 \oplus H_2)^*$ is hence such that $\widetilde{(\rho, \sigma)}(a) \neq 0$. This completes the induction step and the proof of Lemma 12.6. $\qquad\square$

Theorem 12.5 and Corollary 12.3.2 imply

Corollary 12.7 *Let H be a finite abelian group. Then $Q(H) = \mathbb{Q}[H]$.*

Let us analyze further the splitting $\mathbb{Q}[H] = \oplus_{i=1}^k \mathbb{F}_{m_{\sigma_i}}$, where H is a finite abelian group and $\sigma_1, \ldots, \sigma_k$ are representatives of the equivalence classes in H^*. We may assume that σ_1 is the trivial character $H \to \{1\} \subset S^1$. It extends to the augmentation

$$\tilde{\sigma}_1 = \text{aug} \colon \mathbb{Q}[H] \to \mathbb{Q} \subset \mathbb{C}, \qquad \sum_{h \in H} q_h h \mapsto \sum_{h \in H} q_h.$$

The trivial character σ_1 is the only element in its equivalence class. Clearly, $m_{\sigma_1} = 1$ and $\mathbb{F}_{m_{\sigma_1}} = \mathbb{Q}$.

The ideal $I(H) = \mathrm{Ker}(\mathrm{aug}\colon \mathbb{Q}[H] \to \mathbb{Q}) \subset \mathbb{Q}[H]$ is called the *augmentation ideal* of $\mathbb{Q}[H]$. Set $\Sigma = \sum_{h \in H} h \in \mathbb{Q}[H]$. The map

$$s\colon \mathbb{Q} \to \mathbb{Q}[H], \qquad q \mapsto \frac{q}{|H|}\Sigma$$

is a ring homomorphism and a section of aug. Therefore the short exact sequence

$$0 \to I(H) \hookrightarrow \mathbb{Q}[H] \xrightarrow{\mathrm{aug}} \mathbb{Q} \to 0$$

splits: $\mathbb{Q}[H] = s(\mathbb{Q}) \oplus I(H)$. If $\sigma \in H^*$ is a non-trivial character, then $\sigma(H)$ is a non-trivial finite cyclic subgroup of S^1 and $\tilde{\sigma}(\Sigma) = \sum_{h \in H} \sigma(h) = 0$. We conclude that the ring isomorphism $\oplus_{i=1}^{k} \tilde{\sigma}_i\colon \mathbb{Q}[H] \to \oplus_{i=1}^{k} \mathbb{F}_{m_{\sigma_i}}$ is the direct sum of the ring isomorphisms

$$\tilde{\sigma}_{1|s(\mathbb{Q})}\colon s(\mathbb{Q}) \to \mathbb{Q} \quad \text{and} \quad \bigoplus_{i=2}^{k} \tilde{\sigma}_{i|I(H)}\colon I(H) \to \bigoplus_{i=2}^{k} \mathbb{F}_{m_{\sigma_i}}.$$

Remark 12.8 Observe that the augmentation ideal $I(H)$ is an algebra with unit $1 - \frac{1}{|H|}\Sigma$. If H is a cyclic group of order n and h is a generator of H, then $h - 1 \in I(H)$, and the identity

$$(h-1)(1 + 2h + 3h^2 + \cdots + nh^{n-1}) = n - \Sigma = n\left(1 - \frac{1}{|H|}\Sigma\right) \in I(H)$$

shows that $h - 1$ is invertible in $I(H)$.

Theorem 12.9 *Let H be a finitely generated infinite abelian group. Then $\mathbb{Q}[H]$ splits in a unique way as a direct sum of finitely many domains, $\mathbb{Q}[H] = \oplus_{i=1}^{k} R_i$.*

Proof. Set $G = H/\mathrm{Tors}\, H$. Let $\mathbb{Q}[\mathrm{Tors}\, H] = \oplus_{i=1}^{k} \mathbb{F}_{m_{\sigma_i}}$ be the splitting of $\mathbb{Q}[\mathrm{Tors}\, H]$ provided by Theorem 12.5. Since G is a free abelian group, the short exact sequence $0 \to \mathrm{Tors}\, H \to H \to G \to 0$ splits and therefore $H = \mathrm{Tors}\, H \oplus G$. Hence,

$$\begin{aligned} \mathbb{Q}[H] &= \mathbb{Q}[\mathrm{Tors}\, H \oplus G] \\ &= (\mathbb{Q}[\mathrm{Tors}\, H])[G] \\ &= \left(\bigoplus_{i=1}^{k} \mathbb{F}_{m_{\sigma_i}}\right)[G] = \bigoplus_{i=1}^{k} \mathbb{F}_{m_{\sigma_i}}[G]. \end{aligned}$$

The rings $R_i = \mathbb{F}_{m_{\sigma_i}}[G]$ are domains. Observe that the splitting $H = \operatorname{Tors} H \oplus G$ is not unique. Nevertheless, by Lemma 12.1, the splitting $\mathbb{Q}[H] = \oplus_{i=1}^{k} R_i$ is unique. $\qquad\square$

Corollary 12.10 *Let H be a finitely generated abelian group. Then $Q(H) = \oplus_{i=1}^{k} Q(R_i)$ is a direct sum of fields.*

The next lemma will be used in Section 13.

Lemma 12.11 *The projection $H \to G = H/\operatorname{Tors} H$ uniquely extends to a ring homomorphism $pr: Q(H) \to Q(G)$.*

Proof. Extend the projection $H \to G$ to a ring homomorphism $p: \mathbb{Q}[H] \to \mathbb{Q}[G]$. Since $\mathbb{Q}[\operatorname{Tors} H] = \mathbb{Q} \oplus I(\operatorname{Tors} H)$, we have that

$$\mathbb{Q}[H] = (\mathbb{Q}[\operatorname{Tors} H])[G] = \mathbb{Q}[G] \oplus (I(\operatorname{Tors} H))[G].$$

Clearly, p is the projection onto the summand $\mathbb{Q}[G]$. Therefore, p maps non-zero-divisors to non-zero-divisors and so extends to the ring homomorphism $pr: Q(H) \to Q(G)$ by the formula $a\,b^{-1} \mapsto p(a)\,p(b)^{-1}$. The uniqueness of pr is clear. $\qquad\square$

13   The maximal abelian torsion

Let X be a finite connected CW-complex. Since the chain group $C_1(X)$ is finitely generated, the group $H = H_1(X; \mathbb{Z})$ is a finitely generated abelian group. Let $\mathbb{Q}[H] = \oplus_{i=1}^{k} R_i$ be the splitting of $\mathbb{Q}[H]$ into a direct sum of domains (cf. Section 12). Let $\varphi_i: \mathbb{Z}[H] \to Q(R_i)$ be the composition

$$\mathbb{Z}[H] \hookrightarrow \mathbb{Q}[H] \hookrightarrow Q(H) = \bigoplus_{i=1}^{k} Q(R_i) \xrightarrow{p_i} Q(R_i),$$

where p_i is the projection onto the i-th summand. If $H_*^{\varphi_i}(X) = 0$, then the torsion $\tau_{\varphi_i}(X) \in (Q(R_i))^* / \pm \varphi_i(H)$ is defined. Recall from Section 6 that $\tau_{\varphi_i}(X)$ is the image of an element of $(Q(R_i))^*$ which depends on the choice of an order, orientations and lifts of the cells of X to the maximal abelian covering $\widehat{X}$ of X. We now require that these choices are the same for all $i \in \{1, \dots, k\}$. Given an ordered oriented family of lifts $\{\hat{e}\}$, set

$$\tau_i(X, \hat{e}) = \begin{cases} \tau_{\varphi_i}(X, \hat{e}) & \in Q(R_i) & \text{if} & H_*^{\varphi_i}(X) = 0, \\ 0 & \in Q(R_i) & \text{if} & H_*^{\varphi_i}(X) \neq 0, \end{cases} \qquad (13.1)$$

and

$$\tau(X, \hat{e}) = \sum_{i=1}^{k} \tau_i(X, \hat{e}) \in \bigoplus_{i=1}^{k} Q(R_i) = Q(H). \qquad (13.2)$$

If $\{\hat{e}'\}$ is another ordered oriented family of lifts, then $\tau(X, \hat{e}') = \pm h \tau(X, \hat{e})$ for some $h \in H$. Thus, the torsion $\tau(X) = \pm \tau(X, \hat{e}) \cdot H \in Q(H)/\pm H$ is a well-defined invariant of X. This torsion is preserved under subdivision of X. It was introduced in [33]. We call $\tau(X)$ the *maximal abelian torsion* of X.

In the case where the group $H = H_1(X; \mathbb{Z})$ is finite we have $\tau(X) \in Q(H)/\pm H = \mathbb{Q}[H]/\pm H$. Therefore the sum of coefficients of $\tau(X)$ is well defined, at least up to sign. We claim that this sum equals 0, so that $\tau(X)$ belongs to the augmentation ideal $I(H) = \mathrm{Ker}(\mathrm{aug} \colon \mathbb{Q}[H] \to \mathbb{Q})$. Indeed, by Theorem 12.5, $\mathbb{Q}[H] = \oplus_{i=1}^{k} \mathbb{F}_{m_{\sigma_i}}$. The map

$$\varphi_i \colon \mathbb{Z}[H] \hookrightarrow \mathbb{Q}[H] \xrightarrow{\oplus_{i=1}^{k} \tilde{\sigma}_i} \bigoplus_{i=1}^{k} \mathbb{F}_{m_{\sigma_i}} \xrightarrow{p_i} \mathbb{F}_{m_{\sigma_i}}$$

equals $\tilde{\sigma}_i$. If σ_1 is the trivial character of H, then

$$C^{\varphi_1}(X) = \mathbb{Q} \otimes_{\varphi_1} C(\hat{X}) = C(X; \mathbb{Q}).$$

Since $H_0(X; \mathbb{Q}) = \mathbb{Q}$, the complex $C^{\varphi_1}(X)$ is not acyclic. Hence $\tau_1(X, \hat{e}) = 0$ and so $\tau(X) \in I(H)/\pm H$.

Example 13.1 Let $X = L = L(p; q_1, \ldots, q_n)$, $p \geq 2$, be a lens space. Let T be the (distinguished) generator of the cyclic group $\pi_1(L) = H_1(L) = H$ of order p. Define $r_i \in \mathbb{Z}/p\mathbb{Z}$ by $r_i q_i \equiv 1 \pmod{p}$ for $i = 1, \ldots, n$. Theorem 10.6 implies that for any non-trivial character σ of H,

$$\tau_{\tilde{\sigma}}(L, \hat{e}) = \pm t^d \prod_{i=1}^{n} (t^{r_i} - 1)^{-1} \in \mathbb{F}_{m_\sigma}$$

where $t = \sigma(T)$ and $d \in \mathbb{Z}/p\mathbb{Z}$. (Observe that the proof of Theorem 10.6 goes through for $p = 2$.) The proof of Theorem 10.6 shows that the sign $\pm$ in this formula and the residue d do not depend on the choice of $\sigma \neq 1$. Hence

$$\tau(L) = \prod_{i=1}^{n} (T^{r_i} - 1)^{-1} \in I(H)/\pm H.$$

Notice that, by Remark 12.8, $T^{r_i} - 1$ is invertible in $I(H)$ for all i.   ◇

The next lemma shows that the maximal abelian torsion $\tau(X)$ determines the Milnor torsion $\tau_\mu(X)$.

Lemma 13.2 *Let X be a finite connected CW-complex. Set $H = H_1(X)$ and $G = H/\operatorname{Tors} H$. Let $pr: Q(H) \to Q(G)$ be the projection provided by Lemma 12.11. Then*

$$pr(\tau(X)) = \tau_\mu(X).$$

Proof. Set $I = I(\operatorname{Tors} H) \subset \mathbb{Q}[\operatorname{Tors} H]$. Recall from the proof of Lemma 12.11 that pr is induced by the projection $\mathbb{Q}[H] = \mathbb{Q}[G] \oplus I[G] \to \mathbb{Q}[G]$. Hence, pr is the projection $Q(H) = Q(G) \oplus Q(I[G]) \to Q(G)$.

Let

$$\tau(X, \hat{e}) = \tau_1(X, \hat{e}) + \sum_{i=2}^{k} \tau_i(X, \hat{e}) \in Q(G) \oplus Q(I[G]) = Q(H)$$

be a representative of $\tau(X)$. Then $pr(\tau(X, \hat{e})) = \tau_1(X, \hat{e})$. The lemma now follows from the fact that the homomorphisms φ_1, $\mu: \mathbb{Z}[H] \to Q(G)$, which determine $\tau_1(X, \hat{e})$, $\tau_\mu(X)$, are equal. $\square$

Lemma 13.2 admits the following generalization which shows that $\tau(X)$ determines all non-zero Reidemeister torsions of X arising from fields of characteristic 0.

Theorem 13.3 *Consider a finite connected CW-complex X. Set $H = H_1(X)$. Let $\mathbb{F}$ be a field and $\varphi: \mathbb{Q}[H] \to \mathbb{F}$ be a ring homomorphism. Set*

$$Q_\varphi = Q_\varphi(\mathbb{Q}[H]) \subset Q(H)$$

(cf. Section 12.1). By Lemma 12.4, φ uniquely extends to a ring homomorphism $\tilde{\varphi}: Q_\varphi \to \mathbb{F}$. If the torsion $\tau_\varphi(X)$ (corresponding to the restriction of φ to $\mathbb{Z}[H]$) does not vanish, then $\tau(X) \in Q_\varphi/\pm H$, and

$$\tau_\varphi(X) = \tilde{\varphi}(\tau(X)) \in \mathbb{F}^*/\pm \varphi(H).$$

Proof. Let $\mathbb{Q}[H] = \oplus_{i=1}^{k} R_i$ be the splitting of $\mathbb{Q}[H]$ as a direct sum of domains. For $i \neq j$, we have $\varphi(R_i) \cdot \varphi(R_j) = \varphi(R_i \cdot R_j) = \varphi(0) = 0$. Since $\mathbb{F}$ is a field, $\varphi(R_i) = 0$ except for one $i_0 \in \{1, \ldots, k\}$. The map φ is thus the composition

$$\mathbb{Q}[H] = \bigoplus_{i=1}^{k} R_i \xrightarrow{p_{i_0}} R_{i_0} \xrightarrow{\varphi|R_{i_0}} \mathbb{F},$$

where p_{i_0} is the projection onto R_{i_0}. Therefore

$$Q_\varphi = Q_{\varphi | R_{i_0}}(R_{i_0}) \oplus \bigoplus_{i \neq i_0} Q(R_i) \subset \bigoplus_i Q(R_i) = Q(H).$$

Let $\tau(X, \hat{e}) = \sum_{i=1}^k \tau_i(X, \hat{e}) \in \oplus_{i=1}^k Q(R_i)$ be a representative of $\tau(X)$, and let $\tau_\varphi(X, \hat{e}) \in \mathbb{F}^*$ be the representative of $\tau_\varphi(X)$ corresponding to the same lift $\{\hat{e}\}$ of the ordered oriented cells of X. Write $\hat{\tau}_i = \tau_i(X, \hat{e})$.

Claim 13.4
  1. $\hat{\tau}_{i_0} \in Q_{\varphi | R_{i_0}}(R_{i_0}) \subset Q_\varphi$.
  2. $\tilde{\varphi}(\hat{\tau}_{i_0}) = \tau_\varphi(X, \hat{e})$.

The claim implies the theorem. Indeed, pick non-zero-divisors $y_i \in R_i$ such that $\hat{\tau}_i y_i \in R_i$. By the first part of the claim, y_{i_0} can be chosen so that $\varphi(y_{i_0}) \neq 0$. Thus,

$$\varphi\left(\sum_i y_i\right) = \varphi(y_{i_0}) \neq 0. \tag{13.3}$$

Observe that $(\sum_i \hat{\tau}_i)(\sum_i y_i) = \sum_i \hat{\tau}_i y_i \in \oplus_i R_i = \mathbb{Q}[H]$. Hence,

$$\tau(X, \hat{e}) = \sum_i \hat{\tau}_i = \left(\sum_i \hat{\tau}_i y_i\right)\left(\sum_i y_i\right)^{-1}. \tag{13.4}$$

Equations (13.3) and (13.4) imply that $\tau(X, \hat{e}) \in Q_\varphi$. Moreover, by (13.4) and Claim 13.4.2,

$$\begin{aligned} \tilde{\varphi}(\tau(X, \hat{e})) &= \left(\sum_i \varphi(\hat{\tau}_i y_i)\right)\left(\sum_i \varphi(y_i)\right)^{-1} \\ &= \varphi(\hat{\tau}_{i_0} y_{i_0})\, \varphi(y_{i_0})^{-1} = \tilde{\varphi}(\hat{\tau}_{i_0}) = \tau_\varphi(X, \hat{e}). \end{aligned}$$

To prove Claim 13.4 we need the following purely algebraic lemma.

Lemma 13.5 Let $\psi: R \to \mathbb{F}$ be a ring homomorphism from a domain R into a field $\mathbb{F}$. Given a based chain complex

$$C = (0 \to C_m \to \cdots \to C_0 \to 0)$$

over R, let C^ψ be the based chain complex $\mathbb{F} \otimes_\psi C$. If C^ψ is acyclic, then so is $Q(R) \otimes_R C$. Moreover, $\tau(Q(R) \otimes_R C) \in Q_\psi(R)$, and

$$\tau(C^\psi) = \tilde{\psi}\left(\tau(Q(R) \otimes_R C)\right).$$

Let us deduce Claim 13.4 from this lemma. Set $R = R_{i_0}$, $\psi = \varphi_{|R_{i_0}} : R \to \mathbb{F}$ and $C = R_{i_0} \otimes_{p_{i_0}} C(\widehat{X})$, where $\widehat{X}$ is the maximal abelian covering of X and C is based by $\hat{e}$. Then

$$C^\psi = \mathbb{F} \otimes_{\varphi_{|R_{i_0}}} C = \mathbb{F} \otimes_{\varphi_{|R_{i_0}}} \left(R_{i_0} \otimes_{p_{i_0}} C(\widehat{X}) \right) = \mathbb{F} \otimes_\varphi C(\widehat{X}) = C^\varphi(X).$$

Since $\tau_\varphi(X) \neq 0$, this complex is acyclic, and so, by the lemma, the chain complex

$$Q(R_{i_0}) \otimes_{R_{i_0}} C = Q(R_{i_0}) \otimes_{R_{i_0}} \left(R_{i_0} \otimes_{p_{i_0}} C(\widehat{X}) \right) = Q(R_{i_0}) \otimes_{p_{i_0}} C(\widehat{X})$$

is acyclic,

$$\hat{\tau}_{i_0} = \tau(Q(R_{i_0}) \otimes_{p_{i_0}} C(\widehat{X})) \in Q_{\varphi_{|R_{i_0}}}(R_{i_0})$$

and

$$\tilde{\varphi}(\hat{\tau}_{i_0}) = \tilde{\psi}(\hat{\tau}_{i_0}) = \tau(C^\varphi(X)) = \tau_\varphi(X, \hat{e}).$$

Proof of Lemma 13.5. Given a matrix $S = (s_{ij})$ over R, set $S^\psi = (\psi(s_{ij}))$. Recall from Section 2 that a τ-chain for C is a sequence $(\alpha_0 = \emptyset, \alpha_1, \ldots, \alpha_m)$ of subsets α_i of the distinguished basis of C_i such that the associated matrices S_i are square matrices. If $(\{\alpha_i\}, \{S_i\})$ is a τ-chain for C, then $(\{\alpha_i\}, \{S_i^\psi\})$ is a τ-chain for C^ψ. Clearly, every τ-chain for C^ψ is obtained in this way. By assumption, C^ψ is acyclic. Then, by Lemma 2.5, there is a non-degenerate τ-chain $(\{\alpha_i\}, \{S_i^\psi\})$ for C^ψ so that $\det S_i^\psi \neq 0$ for all i. This τ-chain lifts to a τ-chain $(\{\alpha_i\}, \{S_i\})$ for C. Let $\iota : R \hookrightarrow Q(R)$ be the inclusion of R into its quotient field. Since $0 \neq \det S_i^\psi = \psi(\det S_i)$, we have $\det S_i \neq 0$, and so $\det S_i^\iota \neq 0$, $i = 1, \ldots, m$. Hence, the τ-chain $(\{\alpha_i\}, \{S_i^\iota\})$ for $Q(R) \otimes_R C$ is non-degenerate. Lemma 2.5 implies that $Q(R) \otimes_R C$ is acyclic. By Theorem 2.2,

$$\tau(Q(R) \otimes_R C) = \pm \prod_{i=1}^m (\det S_i^\iota)^{(-1)^{i+1}}$$

$$= \pm \left(\prod_{i \text{ odd}} \det S_i \right) \left(\prod_{i \text{ even}} \det S_i \right)^{-1} \in Q_\psi(R).$$

Finally,

$$\tilde{\psi}\left(\tau(Q(R) \otimes_R C) \right) = \pm \prod_{i=1}^m \left(\det S_i^\psi \right)^{(-1)^{i+1}} = \tau(C^\psi).$$

This completes the proof of the lemma and of Theorem 13.3. $\qquad\qquad\square$

Remarks 13.6

   1. For 2-dimensional CW-complexes and closed 3-manifolds, Theorem 13.3 also holds in the case $\tau_\varphi(X) = 0$ provided that $\varphi(H) \neq 1$.

   2. Let $in \colon \mathbb{Z}[H] \hookrightarrow Q(H)$ be the inclusion. The torsion $\tau_{in}(X) \in Q(H)$ is defined if and only if $H_*^{\varphi_i}(X) = 0$ for all $i = 1, \dots, k$. If it is defined, then $\tau_{in}(X) = \tau(X)$.

Proof. Since

$$
\begin{aligned}
C^{in}(X) &= Q(H) \otimes_{in} C(\widehat{X}) \\
&= \left(\bigoplus_{i=1}^{k} Q(R_i) \right) \otimes_{in} C(\widehat{X}) \\
&= \bigoplus_{i=1}^{k} \left(Q(R_i) \otimes_{\varphi_i} C(\widehat{X}) \right) \\
&= \bigoplus_{i=1}^{k} C^{\varphi_i}(X),
\end{aligned}
$$

the chain complex $C^{in}(X)$ is acyclic if and only if $H_*^{\varphi_i}(X) = 0$ for all $i = 1, \dots, k$. The last statement follows from definitions. □

14   Torsions of manifolds

A *triangulation* of a topological space M is a pair (X, t), where X is a simplicial complex and t is a homeomorphism from its underlying topological space $|X|$ to M. A triangulation (X', t') is a *linear subdivision* of (X, t) if $t^{-1} \circ t' \colon |X'| \to |X|$ maps each simplex of X' linearly into a simplex of X. Let $x \in X$ be a vertex. The *star of x* is the union of closed simplices a containing x,

$$
\operatorname{st}_X(x) = \bigcup_{x \in a \in X} a.
$$

A *piecewise-linear manifold* (*pl-manifold*) of dimension m is an m-dimensional topological manifold M endowed with a maximal family of triangulations, called *pl-triangulations*, such that

- any two *pl*-triangulations have a common linear subdivision which is *pl*,

- for any point $x \in M$ there exists a *pl*-triangulation X of M such that x is a vertex of X and $\operatorname{st}_X(x)$ is simplicially isomorphic to a linear triangulation of a closed m-dimensional simplex.

If $\partial M \neq \emptyset$, then ∂M inherits the structure of a *pl*-manifold from M.

By a theorem of Whitehead [38, 26], each compact smooth (C^∞-) manifold
has a canonical pl-structure, unique up to ambient isotopy. This pl-structure
contains a C^1-triangulation. In dimensions 2 and 3, each topological manifold
carries a pl-structure and a smooth structure, both unique up to ambient isotopy. In dimension 4, each compact pl-manifold has a unique smooth structure,
but there are closed topological manifolds which admit no pl-structure.

Let M be a compact connected pl-manifold. Let Λ be a ring as in Section
3.2 and $\varphi\colon \mathbb{Z}[\pi] \to \Lambda$ be a ring homomorphism, where $\pi = \pi_1(M)$. Assume
that $H^\varphi_*(M) = 0$. Endow M with the CW-decomposition induced by any pl-
triangulation of M. By the invariance of torsions under subdivision, the torsion
$\tau_\varphi(M) \in K_1(\Lambda)/\pm\varphi(\pi)$ is independent of the choice of triangulation in the
pl-class. In particular, we have the Milnor torsion $\tau_\mu(M) \in Q(G)/\pm G$ and
the maximal abelian torsion $\tau(M) \in Q(H)/\pm H$, where $H = H_1(M)$ and
$G = H/\operatorname{Tors} H$. We can similarly use pl-triangulations of M to define the
torsions of the pair $(M, \partial M)$.

Theorem 14.1 [15, 23] *Let M be a compact, connected pl-manifold of dimen-
sion m. Set $\pi = \pi_1(M)$. Let Λ be a ring as in Section 3.2, and let $\varphi\colon \mathbb{Z}[\pi] \to \Lambda$
be a ring homomorphism such that $H^\varphi_*(M) = 0$. Let $\sigma\colon \mathbb{Z}[\pi] \to \mathbb{Z}[\pi]$ be the
involution defined by $\sigma(\alpha) = (-1)^{w_1(\alpha)}\alpha^{-1}$, $\alpha \in \pi$, where $w_1\colon \pi \to \mathbb{Z}/2\mathbb{Z}$ is the
first Stiefel–Whitney class of M (see Section 5.5). Then $H^{\varphi\circ\sigma}_*(M, \partial M) = 0$
and*

$$\tau_{\varphi\circ\sigma}(M, \partial M) = \tau_\varphi(M)^{(-1)^{m+1}} \in K_1(\Lambda)/\pm\varphi(\pi).$$

In particular, if $\partial M = \emptyset$, then $H^{\varphi\circ\sigma}_*(M) = 0$ and

$$\tau_{\varphi\circ\sigma}(M) = \tau_\varphi(M)^{(-1)^{m+1}}.$$

Corollary 14.2 *Suppose in addition that the ring Λ is endowed with an involution $\lambda \mapsto \bar{\lambda}$ such that*

$$\overline{\varphi(\alpha)} = (-1)^{w_1(\alpha)}\varphi(\alpha^{-1}) \qquad \text{for all } \alpha \in \pi.$$

Denote the induced involution in $K_1(\Lambda)$ also by $\bar{\ }$. Then

$$\tau_\varphi(M, \partial M) = \overline{\tau_\varphi(M)}^{(-1)^{m+1}} \in K_1(\Lambda)/\pm\varphi(\pi).$$

Proof. Since $\varphi \circ \sigma = \bar{\varphi}$ holds on π, it holds on $\mathbb{Z}[\pi]$. We have

$$\tau_{\varphi\circ\sigma}(M, \partial M) = \tau_{\bar{\varphi}}(M, \partial M) = \overline{\tau_\varphi(M, \partial M)}$$

where the last equality follows from the obvious functoriality of the torsion of a chain complex with respect to isomorphisms of the ground ring. The claim now follows from Theorem 14.1.   □

Proof of Theorem 14.1. The proof is based on the same ideas as the classical proof of the Poincaré duality. We start with recalling some generalities on triangulations. Given a simplex a, let $\underline{a}$ be its barycenter. If a is a face of a simplex b, write $a \leq b$. If a is a proper face of b, i.e., $a \leq b$ and $a \neq b$, write $a < b$. Given an increasing sequence of simplices $a_0 < a_1 < \cdots < a_k$, let

$$\langle \underline{a}_0, \underline{a}_1, \ldots, \underline{a}_k \rangle$$

be the convex hull of the points $\underline{a}_0, \underline{a}_1, \ldots, \underline{a}_k$. This is a simplex of dimension k contained in a_k (cf. Figure 14.1).

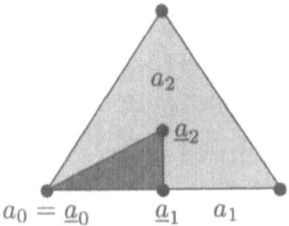

Figure 14.1: $\langle \underline{a}_0, \underline{a}_1, \underline{a}_2 \rangle$.

Fix a simplex a and consider the simplices $\langle \underline{a}_0, \underline{a}_1, \ldots, \underline{a}_k \rangle$ corresponding to all increasing chains

$$a_0 < a_1 < \cdots < a_k \leq a, \qquad k = 0, 1, \ldots, \dim a.$$

These simplices form a triangulation of a. It is called the *first barycentric subdivision* of a (cf. Figure 14.2).

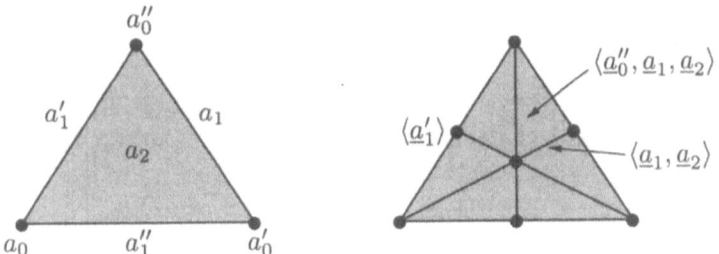

Figure 14.2: a and its first barycentric subdivision.

Let X be a triangulated space. Applying the first barycentric subdivision to all simplices of X, we obtain a new triangulation of X, called its *first barycentric subdivision* and denoted by $X^{(1)}$. The simplices of $X^{(1)}$ have the form $\langle \underline{a}_0, \underline{a}_1, \ldots, \underline{a}_k \rangle$, where $a_0 < a_1 < \cdots < a_k$ is an increasing sequence of simplices of X.

We first prove Theorem 14.1 in the case $\partial M = \emptyset$. Fix a *pl*-triangulation X of M. Let a be an n-dimensional closed simplex of X. The set

$$a^* = \bigcup_{a=a_0<a_1<\cdots<a_k,\, k\geq 0} \langle \underline{a}_0, \underline{a}_1, \ldots, \underline{a}_k \rangle \qquad (14.1)$$

is called the *dual cell* of a. Set

$$\partial a^* = \bigcup_{a<a_1<\cdots<a_k,\, k\geq 1} \langle \underline{a}_1, \ldots, \underline{a}_k \rangle \subset a^* \qquad (14.2)$$

and $\operatorname{Int} a^* = a^* \setminus \partial a^*$.

Example 14.3 In Figure 14.3, X is 2-dimensional and formed by six 2-simplices. The first barycentric subdivision $X^{(1)}$ is drawn with dashed lines. In the left figure, a is a vertex of X and a^* is the shaded region. In the right figure, a is a 1-simplex.

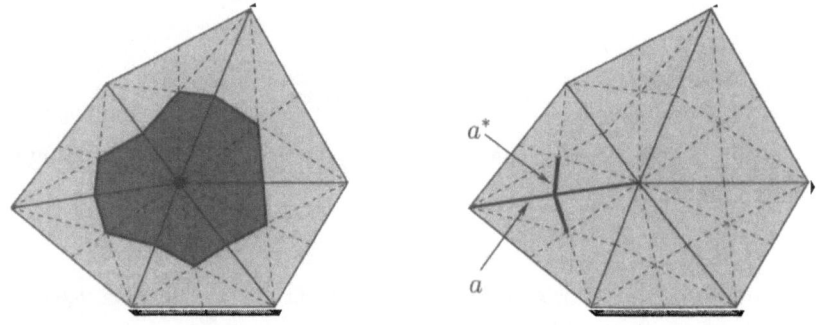

Figure 14.3: The dual cell of a vertex and of a 1-simplex.

Since X is locally *pl*-isomorphic to a closed m-simplex, for each n-simplex a the pair $(a^*, \partial a^*)$ is homeomorphic to (D^{m-n}, S^{m-n-1}) (see [30, Chapter 2]). In particular, the dual cell a^* is an $(m-n)$-cell. When a runs over all simplices of X, the dual cells a^* form a CW-decomposition of M. It is called the *dual cellular decomposition* of X and denoted by X^*. Observe that $X^{(1)}$ is a CW-subdivision of both X and X^*.

If $b < a$, then every sequence $a = a_0 < a_1 < \cdots < a_k$ can be extended to $b < a_0 < a_1 < \cdots < a_k$, and so $\langle \underline{a}_0, \ldots, \underline{a}_k \rangle \subset \langle \underline{b}, \underline{a}_0, \ldots, \underline{a}_k \rangle$, whence $a^* \subset \partial b^*$. The equality $\operatorname{codim}_b a = \operatorname{codim}_{a^*} b^*$ implies that if $\operatorname{codim}_b a = 1$, then a^* is a cell of maximal dimension in the sphere ∂b^*.

Let us lift the triangulation X and the dual cellular decomposition X^* of M to a triangulation $\widetilde{X}$ and a CW-decomposition $\widetilde{X}^*$ of the universal covering $\widetilde{M}$ of M. Clearly, $\widetilde{X}^*$ is dual to $\widetilde{X}$. As in Section 6.1 we orient and order all simplices $a \in X$ and choose over each simplex $a \in X$ a lift $\tilde{a} \in \widetilde{X}$ which is oriented in such a way that the projection $\tilde{a} \to a$ is orientation preserving. In this way, $C(\widetilde{X})$ becomes a free chain complex over $\mathbb{Z}[\pi]$ with basis $\tilde{e} = \{\tilde{a}\}$. The dual cells $\tilde{e}^* = \{\tilde{a}^*\}$ form an ordered set of cells of $\widetilde{X}^*$ such that over each cell of X^* lies exactly one cell of $\tilde{e}^*$. We orient each cell $\tilde{a}^*$ as follows. Since the underlying space $\widetilde{M}$ of $\widetilde{X}$ is a simply-connected manifold, it is orientable. Fix one of the two possible orientations of $\widetilde{M}$. We orient $\tilde{a}^*$ in such a way that the orientation of $\widetilde{M}$ at $\tilde{a} \cap \tilde{a}^*$ given by the orientation of $\tilde{a}$ followed by the orientation of $\tilde{a}^*$ coincides with the given one. In this way the set $\tilde{e}^*$ becomes a basis of the chain complex $C(\widetilde{X}^*)$ over $\mathbb{Z}[\pi]$.

Specifically, let $\{a_l^i\}_l$ be the (ordered and oriented) i-simplices of X. Let $\tilde{e}_i = \{\tilde{a}_l^i\}_l$ be the corresponding basis of $C_i(\widetilde{X})$ and $\tilde{e}_i^* = \{(\tilde{a}_l^i)^*\}_l$ be the basis of $C_{m-i}(\widetilde{X}^*)$ determined by the dual cells. Let A_i be the matrix of $\partial_i \colon C_{i+1}(\widetilde{X}) \to C_i(\widetilde{X})$ with respect to $\tilde{e}_{i+1}, \tilde{e}_i$ and let B_{m-i-1} be the matrix of $\partial_{m-i-1} \colon C_{m-i}(\widetilde{X}^*) \to C_{m-i-1}(\widetilde{X}^*)$ with respect to $\tilde{e}_i^*, \tilde{e}_{i+1}^*$. Given a matrix $S = (s_{kl})$ over $\mathbb{Z}[\pi]$, set $S^\sigma = (\sigma(s_{kl}))$ and $S^T = (s_{lk})$.

Claim 14.4 $B_{m-i-1} = (-1)^{i+1}(A_i^\sigma)^T$ for all i.

This claim implies Theorem 14.1 for $\partial M = \emptyset$. Indeed, the bases $\tilde{e}$ and $\tilde{e}^*$ of $C(\widetilde{X})$ and $C(\widetilde{X}^*)$ yield bases of the chain complexes $C^\varphi(X) = \Lambda \otimes_\varphi C(\widetilde{X})$ and $C^{\varphi\sigma}(X^*) = \Lambda \otimes_{\varphi\sigma} C(\widetilde{X}^*)$, respectively. With respect to these bases, the boundary homomorphisms of $C^\varphi(X)$ and $C^{\varphi\sigma}(X^*)$ are given by the matrices A_i^φ and $B_{m-i-1}^{\varphi\sigma}$. By Claim 14.4,

$$B_{m-i-1}^{\varphi\sigma} = (-1)^{i+1}((A_i^\sigma)^T)^{\varphi\sigma} = (-1)^{i+1}(A_i^\varphi)^T.$$

For all i, we can identify $C_{m-i}^{\varphi\sigma}(X^*)$ with $(C_i^\varphi(X))^* = \mathrm{Hom}_\Lambda(C_i^\varphi(X), \Lambda)$ in such a way that the basis $\tilde{e}_i^*$ of $C_{m-i}^{\varphi\sigma}(X^*)$ is dual to the basis $\tilde{e}_i$ of $C_i^\varphi(X)$. By the definition of the dual chain complex, the complex $C^{\varphi\sigma}(X^*)$ is thus the dual chain complex of $C^\varphi(X)$. Since $C^\varphi(X)$ is acyclic, so is $C^{\varphi\sigma}(X^*)$. Theorem 3.4 implies that $\tau_{\varphi\sigma}(X^*, \tilde{e}^*) = \pm \tau_\varphi(X, \tilde{e})^{(-1)^{m+1}}$. Theorem 14.1 now follows from Theorem 6.1 and the fact that the pl-triangulation $X^{(1)}$ of M is a common subdivision of X and X^*.

Proof of Claim 14.4. The matrix $A_i = (\lambda_{k,l})$ is given by

$$\partial \tilde{a}_k^{i+1} = \sum_l \lambda_{k,l} \tilde{a}_l^i,$$

where $\partial = \partial_i \colon C_{i+1}(\tilde{X}) \to C_i(\tilde{X})$ and $\lambda_{k,l} \in \mathbb{Z}[\pi]$ for all k, l. If a_l^i is not a face of a_k^{i+1}, then $\lambda_{k,l} = 0$. If a_l^i is a face of a_k^{i+1}, then $\lambda_{k,l} = \varepsilon_{k,l}\, h_{k,l}$, where $\varepsilon_{k,l} = \pm 1$ and $h_{k,l} \in \pi$ is the unique element of π such that $h_{k,l}\, \tilde{a}_l^i \subset \tilde{a}_k^{i+1}$.

The matrix $B_{m-i-1} = (\mu_{l,k})$ is given by

$$\partial(\tilde{a}_l^i)^* = \sum_k \mu_{l,k}\, (\tilde{a}_k^{i+1})^*$$

where $\partial = \partial_{m-i-1} \colon C_{m-i}(\tilde{X}^*) \to C_{m-i-1}(\tilde{X}^*)$ and $\mu_{l,k} \in \mathbb{Z}[\pi]$ for all k, l. If a_l^i is not a face of a_k^{i+1}, then $\partial(a_l^i)^*$ does not meet $\mathrm{Int}(a_k^{i+1})^*$, and so $\mu_{l,k} = 0$. If a_l^i is a face of a_k^{i+1}, then

$$h_{k,l}\, \partial(\tilde{a}_l^i)^* = \partial(h_{k,l}\, \tilde{a}_l^i)^* \supset (\tilde{a}_k^{i+1})^*$$

so that $\partial(\tilde{a}_l^i)^* \supset h_{k,l}^{-1}\, (\tilde{a}_k^{i+1})^*$. Since the corresponding incidence number is ± 1, we conclude that $\mu_{l,k} = \varepsilon_{k,l}'\, h_{k,l}^{-1}$ with $\varepsilon_{k,l}' = \pm 1$. Therefore,

$$\mu_{l,k} = \varepsilon_{k,l}\, \varepsilon_{k,l}'\, \varepsilon_{k,l}''\, \sigma(\lambda_{k,l})$$

where $\varepsilon_{k,l}'' = (-1)^{w_1(h_{k,l})} = \pm 1$. It remains to prove that $\varepsilon_{k,l}\, \varepsilon_{k,l}'\, \varepsilon_{k,l}'' = (-1)^{i+1}$ for all k, l such that a_l^i is a face of a_k^{i+1}.

Fix i, k, l as above and set $h = h_{k,l} \in \pi$. It follows from definitions that $\varepsilon_{k,l} = +1$ if the orientation of a_k^{i+1} is given by the outward looking vector ν on a_l^i followed by the orientation of a_l^i, and $\varepsilon_{k,l} = -1$ otherwise (cf. Figure 14.4). Clearly, the orientation of a_k^{i+1} is given by $(\varepsilon_{k,l}\, \nu, \alpha_1, \ldots, \alpha_l)$ where $(\alpha_1, \ldots, \alpha_i)$ is a positive basis in the tangent space of a_l^i.

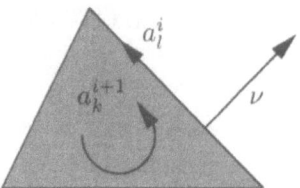

Figure 14.4: An example with $\varepsilon_{k,l} = +1$.

Let $(\beta_1, \ldots, \beta_{m-i-1})$ be a positive basis in the tangent space of the cell $(\tilde{a}_k^{i+1})^*$. By the choice of the orientation in this cell, the fixed orientation of $\widetilde{M}$ is given in the point $\underline{\tilde{a}}_k^{i+1} = \tilde{a}_k^{i+1} \cap (\tilde{a}_k^{i+1})^*$ by the basis of tangent vectors

$$(\varepsilon_{k,l}\, \tilde{\nu}, \tilde{\alpha}_1, \ldots, \tilde{\alpha}_i, \beta_1, \ldots, \beta_{m-i-1})$$

where $\tilde{\nu}, \tilde{\alpha}_1, \ldots, \tilde{\alpha}_i$ are lifts of $\nu, \alpha_1, \ldots, \alpha_i$ to $\tilde{a}_k^{i+1}$. The covering transformation $\widetilde{M} \to \widetilde{M}$ determined by h^{-1} maps the vectors $\tilde{\nu}, \tilde{\alpha}_1, \ldots, \tilde{\alpha}_i, \beta_1, \ldots, \beta_{m-i-1}$

to tangent vectors at the point $h^{-1}\left(\tilde{a}_k^{i+1}\right)$. Denote the latter vectors by $\tilde{\nu}'$, $\tilde{\alpha}_1'$, $\ldots$, $\tilde{\alpha}_i'$, β_1', $\ldots$, β_{m-i-1}', respectively. Note that $h^{-1}\colon \widetilde{M} \to \widetilde{M}$ preserves the orientation if and only if $w_1(h) = 0$. Therefore, the basis of tangent vectors at $h^{-1}\left(\tilde{a}_k^{i+1}\right)$

$$\left(\varepsilon_{k,l}\, \varepsilon_{k,l}''\, \tilde{\nu}', \tilde{\alpha}_1', \ldots, \tilde{\alpha}_i', \beta_1', \ldots, \beta_{m-i-1}'\right)$$

is positive. It is clear that the vectors $(\tilde{\alpha}_1', \ldots, \tilde{\alpha}_i')$ are lifts of $\alpha_1, \ldots, \alpha_i$ to $\tilde{a}_l^i$. Therefore, this sequence of vectors determines the positive orientation in $\tilde{a}_l^i$. Hence, the orientation in $(\tilde{a}_l^i)^*$ is given by $((-1)^i \varepsilon_{k,l}\, \varepsilon_{k,l}''\, \tilde{\nu}', \beta_1', \ldots, \beta_{m-i-1}')$. Observe that the outward looking vector on $h^{-1}(\tilde{a}_k^{i+1})^* \subset (\tilde{a}_l^i)^*$ is $-\tilde{\nu}'$, cf. Figure 14.5. This implies $\varepsilon_{k,l}' = -(-1)^i \varepsilon_{k,l}\, \varepsilon_{k,l}''$. Hence $\varepsilon_{k,l}\, \varepsilon_{k,l}'\, \varepsilon_{k,l}'' = (-1)^{i+1}$.

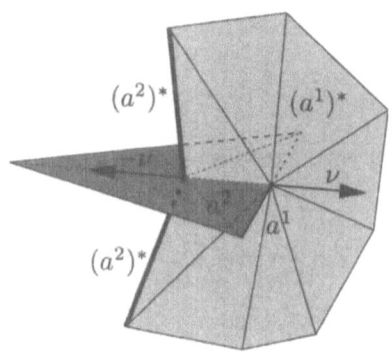

Figure 14.5: A 3-dimensional picture.

Let now $\partial M \neq \emptyset$. Let X be a pl-triangulation of M and let $a \in X$. The dual cell of a^* is again defined by (14.1). The simplices of X lying in ∂M form a pl-triangulation of ∂M, denoted by ∂X. If $a \in \partial X$, we may also consider the dual cell of a in ∂X

$$a_\partial^* = \bigcup_{a=a_0<a_1<\cdots<a_k \in \partial X,\ k\geq 0} \langle a_0, a_1, \ldots, a_k \rangle.$$

Since X and ∂X are pl-triangulations, both a^* and a_∂^* are cells. For $a \in X \backslash \partial X$, define ∂a^* as in (14.2), and for $a \in \partial X$, define

$$\partial a^* = a_\partial^* \cup \bigcup_{a<a_1<\cdots<a_k,\ k\geq 1} \langle a_1, \ldots, a_k \rangle \subset a^*.$$

The cells of both types

$$\{a^* \mid a \in X\} \cup \{a_\partial^* \mid a \in \partial X\}$$

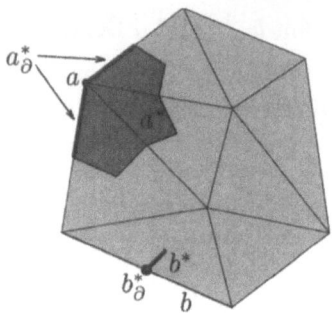

Figure 14.6: The two dual cells of a vertex $a \in \partial X$ and a 1-simplex $b \in \partial X$.

form the *dual cellular decomposition* X^* of M. The CW-complex

$$\partial X^* = \{a_\partial^* \mid a \in \partial X\}$$

is a subcomplex of X^* with underlying space ∂M. Each simplex $a \in \partial X$ contributes two cells to X^*. However, passing to the quotient

$$C(X^*, \partial X^*) = C(X^*)/C(\partial X^*),$$

we obtain a one-to-one correspondence between the basis vectors in $C_i(X)$ and the basis vectors in $C_{m-i}(X^*, \partial X^*)$. As before, the based chain complexes $C^\varphi(X) = \Lambda \otimes_\varphi C(\tilde{X})$ and $C^{\varphi o \sigma}(X^*, \partial X^*) = \Lambda \otimes_{\varphi o \sigma} C(\tilde{X}^*, \partial \tilde{X}^*)$ are dual to each other, and so $H_*^\varphi(X) = 0$ implies that $H_*^{\varphi o \sigma}(X^*, \partial X^*) = 0$. Using the invariance of torsions under cellular subdivisions of CW-pairs, the equation

$$\tau_{\varphi o \sigma}(M, \partial M) = \tau_\varphi(M)^{(-1)^{m+1}}$$

now follows as in the closed case. □

Remark 14.5 Let M be an m-dimensional closed pl-manifold. Since any two pl-triangulations of M have a common pl-subdivision, any quantity ρ computed from a CW-decomposition of M and invariant under cellular subdivision defines an invariant of M. Moreover, to each pl-triangulation X of M we associate the dual cellular decomposition X^* of M. Since $X^{(1)}$ is a common subdivision, $\rho(X) = \rho(X^*)$, and so the geometric duality between X and X^* is reflected in an algebraic duality for ρ.

Consider for instance the Euler characteristic $\chi(X) = \sum_{i=0}^m (-1)^i \alpha_i$, where α_i is the number of i-cells of X. Recall from Section 5.7 that the cellular homology $H_*(M)$ does not depend on the CW-decomposition of M used to compute it. This and (1.1) imply that χ is invariant under cellular subdivision. Writing α'_j for the number of j-cells of X^* and using $\alpha'_j = \alpha_{m-j}$, we find

$$\chi(M) = \chi(X^*) = \sum_{j=0}^{m}(-1)^j \alpha'_j = \sum_{j=0}^{m}(-1)^j \alpha_{m-j} = \sum_{i=0}^{m}(-1)^{m-i}\alpha_i$$

$$= (-1)^m \sum_{i=0}^{m}(-1)^i \alpha_i = (-1)^m \chi(X) = (-1)^m \chi(M).$$

In particular, $\chi(M) = 0$ whenever m is odd. Extending this argument, one obtains the duality for the Betti numbers of M and the Poincaré duality. Passing to CW-pairs and manifolds with boundary, one obtains the Alexander duality and the Lefschetz duality. ◇

We can apply Theorem 14.1 to the Milnor torsion τ_μ and the maximal abelian torsion τ. For simplicity we restrict ourselves to the case of orientable manifolds.

Corollary 14.6 *Let M be a compact connected orientable pl-manifold of dimension m. Let $G = H/\operatorname{Tors} H$, where $H = H_1(M)$, and let $\mu\colon \mathbb{Z}[H] \to Q(G)$ be the natural ring homomorphism (cf. Section 11). Let $q \mapsto \bar{q}$ be the involution in $Q(G)$ sending $g \in G$ to g^{-1}. If m is odd, then*

$$\tau_\mu(M, \partial M) = \overline{\tau_\mu(M)} \in Q(G)/\pm G.$$

If m is even and $\tau_\mu(M) \neq 0$, then

$$\tau_\mu(M, \partial M) = \overline{\tau_\mu(M)}^{-1} \in Q(G)/\pm G.$$

Proof. Set $\pi = \pi_1(M)$. Let $h\colon \mathbb{Z}[\pi] \to \mathbb{Z}[H]$ be the projection induced by the natural projection $\pi \to \pi/[\pi, \pi] = H$ and $\sigma\colon \mathbb{Z}[\pi] \to \mathbb{Z}[\pi]$ be the involution defined by $\sigma(\alpha) = \alpha^{-1}$, $\alpha \in \pi$. Clearly, $\overline{\mu \circ h} = (\mu \circ h) \circ \sigma$.

Let m be odd. If $H_*^{\mu \circ h}(M) \neq 0$, then, by Theorem 14.1, $H_*^{\mu \circ h \circ \sigma}(M, \partial M) \neq 0$. Since $\mu \circ h \circ \sigma = \overline{\mu \circ h}$, we have $H_*^{\mu \circ h}(M, \partial M) \neq 0$. Thus, by definition, $\tau_\mu(M) = \tau_\mu(M, \partial M) = 0$. For $H_*^{\mu \circ h}(M) = 0$, the corollary follows from Corollary 14.2.

If m is even and $\tau_\mu(M) \neq 0$, then $H_*^{\mu \circ h}(M) = 0$, and the claim follows from Corollary 14.2. □

Corollary 14.7 *Let M be a compact connected orientable piecewise-linear 3-manifold whose boundary is non-empty and consists of tori. Then*

$$\overline{\Delta_{\pi_1(M)}} = \Delta_{\pi_1(M)}.$$

Proof. Note that $\overline{t-1} = t^{-1} - 1 = t - 1$ in $\mathbb{Z}[t^{\pm 1}]/\pm\{t^n\}_{n\in\mathbb{Z}}$. The claim now follows from Corollary 11.9, Theorem 11.10 and Corollary 14.6.   □

Corollary 14.8 *Let M be a compact connected orientable pl-manifold of dimension m. Set $H = H_1(M)$. Let $q \mapsto \bar{q}$ be the involution in $Q(H)$ sending $h \in H$ to h^{-1}. If m is odd, then*

$$\tau(M, \partial M) = \overline{\tau(M)} \in Q(H)/\pm H.$$

If m is even and $\tau(M)$ is invertible in $Q(H)$, then

$$\tau(M, \partial M) = \overline{\tau(M)}^{-1} \in Q(H)/\pm H.$$

Proof. As in Section 13, let $Q(H) = \oplus_{i=1}^k Q(R_i)$ be the splitting of $Q(H)$ into a direct sum of fields, and let φ_i be the composition of the inclusion $\mathbb{Z}[H] \hookrightarrow Q(H)$ with the projection $Q(H) \to Q(R_i)$. The involution $Q(H) \to Q(H)$, $q \mapsto \bar{q}$, is induced by $\mathbb{Z}[H] \to \mathbb{Z}[H]$, $h \mapsto h^{-1}$. We denote both involutions by σ. We should prove that

$$\tau(M, \partial M) = \sigma(\tau(M))^{(-1)^{m+1}} \tag{14.3}$$

where in the case of even m we assume that $\tau(M)$ is invertible.

As a ring automorphism, $\sigma: Q(H) \to Q(H)$ maps domains in $Q(H)$ to domains. This and Lemma 12.1 imply that σ permutes the fields $Q(R_i)$. Hence, σ permutes the indices $i = 1, \ldots, k$ so that $Q(R_{\sigma(i)}) = \sigma(Q(R_i))$ for all i. It follows that $\varphi_{\sigma(i)} \circ \sigma = \sigma \circ \varphi_i: \mathbb{Z}[H] \to Q(R_{\sigma(i)})$, and, since σ is an involution,

$$\varphi_i \circ \sigma = \sigma \circ \varphi_{\sigma(i)}: \mathbb{Z}[H] \to Q(R_i).$$

Recall from Section 6.3 that the torsions $\tau(M)$ and $\tau(M, \partial M)$ can be computed on the level of the maximal abelian covering $\widehat{M}$ of M. Since, by Lemma 5.11, $\widehat{M}$ is orientable, we may proceed as in the proof of Theorem 14.1: We choose a pl-triangulation X of M and lift the cells in X to an ordered oriented family $\hat{e}$ of cells in $\widehat{X}$. The family $\hat{e}$ determines an ordered oriented family $\hat{e}^*$ of dual cells in $\widehat{X}^*$. For all i, we compute the torsions $\tau_i(X, \hat{e})$ and $\tau_i(X^*, \partial X^*, \hat{e}^*)$ with respect to these families $\hat{e}$ and $\hat{e}^*$.

Formula (14.3) follows from

Claim 14.9 *For all $i = 1, \ldots k$, we have either*

$$\tau_i(X, \hat{e}) = 0 \quad and \quad \tau_{\sigma(i)}(X^*, \partial X^*, \hat{e}^*) = 0$$

or

$$\sigma(\tau_i(X, \hat{e})) = \pm\tau_{\sigma(i)}(X^*, \partial X^*, \hat{e}^*)^{(-1)^{m+1}} \tag{14.4}$$

where the sign $\pm$ is the same for all i with $\tau_i(X, \hat{e}) \neq 0$.

Note that in the case of even m our assumptions imply that $\tau_i(X, \hat{e}) \neq 0$ for all i.

Let us prove Claim 14.9. Assume first that $\tau_i(X, \hat{e}) = 0$, i.e., $H^{\varphi_i}_*(X) \neq 0$. As in the proof of Theorem 14.1, the chain complexes

$$C^{\varphi_i}(X) \quad \text{and} \quad C^{\varphi_i \circ \sigma}(X^*, \partial X^*) = C^{\sigma \circ \varphi_{\sigma(i)}}(X^*, \partial X^*)$$

are dual to each other, and so $H^{\sigma \circ \varphi_{\sigma(i)}}_*(X^*, \partial X^*) \neq 0$. Since $\sigma \colon Q(R_{\sigma(i)}) \to Q(R_i)$ is an isomorphism, we obtain that $H^{\varphi_{\sigma(i)}}_*(X^*, \partial X^*) \neq 0$. Hence $\tau_{\sigma(i)}(X^*, \partial X^*, \hat{e}^*) = 0$. Assume next that $\tau_i(X, \hat{e}) \neq 0$. We compute

$$\begin{aligned}
\tau_i(X, \hat{e}) &= \tau(C^{\varphi_i}(X, \hat{e})) \\
&= \pm \tau(C^{\varphi_i \circ \sigma}(X^*, \partial X^*, \hat{e}^*))^{(-1)^{m+1}} \\
&= \pm \tau(C^{\sigma \circ \varphi_{\sigma(i)}}(X^*, \partial X^*, \hat{e}^*))^{(-1)^{m+1}} \\
&= \pm \sigma\left(\tau(C^{\varphi_{\sigma(i)}}(X^*, \partial X^*, \hat{e}^*))\right)^{(-1)^{m+1}} \\
&= \pm \sigma\left(\tau_{\sigma(i)}(X^*, \partial X^*, \hat{e}^*)\right)^{(-1)^{m+1}}.
\end{aligned}$$

This implies (14.4). Assume finally that $\tau_i(X, \hat{e}) \neq 0$ and $\tau_{i'}(X, \hat{e}) \neq 0$. Set

$$\begin{aligned}
B_j &= \operatorname{Im}\left(\partial_j \colon C^{\varphi_i}_{j+1}(X, \hat{e}) \to C^{\varphi_i}_j(X, \hat{e})\right), \\
B'_j &= \operatorname{Im}\left(\partial_j \colon C^{\varphi_{i'}}_{j+1}(X, \hat{e}) \to C^{\varphi_{i'}}_j(X, \hat{e})\right).
\end{aligned}$$

By assumption, $C^{\varphi_i}(X, \hat{e})$ and $C^{\varphi_{i'}}(X, \hat{e})$ are acyclic, and so, by (2.1), $\dim B_j = \dim B'_j$ for all j. The independence of the sign $\pm$ of the choice of i now follows from Remark 1.10.2. $\qquad\square$

Our next aim is to relate the Milnor torsion and the Alexander polynomial of a closed 3-manifold.

Lemma 14.10 *Let M be a compact orientable pl-manifold of dimension m. Set $H = H_1(M)$. Define $\sigma \colon \mathbb{Z}[H] \to \mathbb{Z}[H]$ by $h \mapsto h^{-1}$. Let R be a Noetherian unique factorization domain and let $\varphi \colon \mathbb{Z}[H] \to R$ be a ring homomorphism. Then*

$$\operatorname{ord} \operatorname{Tors}_R H^{\varphi}_i(M) = \operatorname{ord} \operatorname{Tors}_R H^{\varphi \circ \sigma}_{m-i-1}(M, \partial M).$$

Proof. We compute $H^{\varphi}_*(M)$ from the based chain complex $C^{\varphi}_*(X)$, where X is a pl-triangulation of M. Let A_i be the matrix of the boundary homomorphism $\partial_i \colon C^{\varphi}_{i+1}(X) \to C^{\varphi}_i(X)$. Let I_i be the ideal of R generated by the minors of A_i of rank $\operatorname{rk} A_i$. By Lemma 4.11,

$$\operatorname{ord} \operatorname{Tors}_R H^{\varphi}_i(M) = \gcd I_i.$$

Let $(X^*, \partial X^*)$ be the dual cellular decomposition of $(M, \partial M)$ induced by the triangulation X. Construct the dual basis of $C_*^{\varphi \circ \sigma}(X^*, \partial X^*)$ as in the proof of Theorem 14.1. Let B_i be the matrix of the boundary homomorphism

$$\partial_i \colon C_{i+1}^{\varphi \circ \sigma}(X^*, \partial X^*) \to C_i^{\varphi \circ \sigma}(X^*, \partial X^*).$$

Then

$$\operatorname{ord} \operatorname{Tors}_R H_i^{\varphi \circ \sigma}(M, \partial M) = \gcd J_i,$$

where J_i denotes the ideal of R generated by the minors of B_i of rank $\operatorname{rk} B_i$. As we have seen in the proof of Theorem 14.1, $(A_i)^T = \pm B_{m-i-1}$. Hence $I_i = J_{m-i-1}$. This implies the claim. $\qquad \square$

Recall from Section 11 that the i'th Alexander module of M is the $\mathbb{Z}[G]$-module $A_i(M) = H_i^g(M)$, where $G = H/\operatorname{Tors} H$ with $H = H_1(M)$ and $g \colon \mathbb{Z}[H] \to \mathbb{Z}[G]$ is the natural projection. In the propositions below we denote by $^-$ and also by σ the canonical involutions in $\mathbb{Z}[H]$ and $\mathbb{Z}[G]$ sending $h \in H$ (resp. $h \in G$) to h^{-1}. Note that $\bar{g} = \sigma \circ g = g \circ \sigma$.

Corollary 14.11 *Let M be a closed connected orientable pl-manifold of dimension m. If $\operatorname{rk} A_i(M) = \operatorname{rk} A_{m-i-1}(M) = 0$, then*

$$\operatorname{ord} A_i(M) = \overline{\operatorname{ord} A_{m-i-1}(M)}.$$

Proof. The assumptions $\operatorname{rk} A_i(M) = \operatorname{rk} A_{m-i-1}(M) = 0$ imply that $A_i(M) = \operatorname{Tors} A_i(M)$ and $A_{m-i-1}(M) = \operatorname{Tors} A_{m-i-1}(M)$. By Lemma 14.10,

$$
\begin{aligned}
\operatorname{ord} A_i(M) &= \operatorname{ord} \operatorname{Tors} H_i^g(M) = \operatorname{ord} \operatorname{Tors} H_{m-i-1}^{g \circ \sigma}(M) \\
&= \operatorname{ord} \operatorname{Tors} H_{m-i-1}^{\sigma \circ g}(M) = \overline{\operatorname{ord} \operatorname{Tors} H_{m-i-1}^g(M)} \\
&= \overline{\operatorname{ord} A_{m-i-1}(M)}. \qquad \square
\end{aligned}
$$

Theorem 14.12 [33] *Let M be a closed connected orientable 3-dimensional pl-manifold with $b_1(M) \geq 1$. Let $\Delta_M = \Delta_{\pi_1(M)} = \operatorname{ord} A_1(M) \in \mathbb{Z}[G]/\pm G$ be the Alexander polynomial of M (where $G = H_1(M)/\operatorname{Tors} H_1(M)$). If $b_1(M) = 1$, let t be a generator of the infinite cyclic group G. Then*

$$\tau_\mu(M) = \begin{cases} \Delta_M (t-1)^{-2} & \text{if} \quad b_1(M) = 1, \\ \Delta_M & \text{if} \quad b_1(M) \geq 2. \end{cases}$$

For a similar result in the non-orientable case, see [33].

Proof of Theorem 14.12. Recall from Section 11 that $\tau_\mu(M) \neq 0$ if and only if $\operatorname{ord} A_i(M) \neq 0$ for all i. Hence, if $\Delta_M = \operatorname{ord} A_1(M) = 0$, then $\tau_\mu(M) = 0$

and the claim follows. Assume that ord $A_1(M) \neq 0$. Then rk $A_1(M) = 0$. Since $b_1(M) \geq 1$, the maximal free abelian covering $\bar{M}$ of M is a non-compact connected 3-manifold. Therefore $A_3(M) = H_3(\bar{M}) = 0$ and rk $A_3(M) = 0$. The $\mathbb{Z}[G]$-module $A_0(M) = H_0(\bar{M}) = \mathbb{Z}$ also has rank 0. Since $0 = \chi(M) = \sum_{i=0}^3 (-1)^i \operatorname{rk} A_i(M)$, we have rk $A_2(M) = 0$. We conclude that rk $A_i(M) = 0$ for all i and therefore ord $A_i(M) \neq 0$ for all i. By Theorem 11.4,

$$\tau_\mu(M) = \prod_{i=0}^3 (\operatorname{ord} A_i(M))^{(-1)^{i+1}} .$$

Clearly, ord $A_3(M) = 1$. Corollary 14.11 and (11.2) imply

$$\operatorname{ord} A_2(M) = \overline{\operatorname{ord} A_0(M)} = \operatorname{ord} A_0(M) \quad = \begin{cases} t-1 & \text{if} \quad b_1(M) = 1, \\ 1 & \text{if} \quad b_1(M) \geq 2. \end{cases}$$

The theorem thus follows. $\qquad\qquad\qquad\qquad\qquad\qquad\qquad\qquad\qquad\qquad\quad\square$

Corollary 14.13 *Let M be as in Theorem 14.12. Then $\overline{\Delta_M} = \Delta_M$.*

First proof. Since in $\mathbb{Z}[G]/\pm G$ we have $\overline{(t-1)^{-2}} = (t-1)^{-2}$, the claim follows from Corollary 14.6 and Theorem 14.12.

Second proof. Since $3 - 1 - 1 = 1$, the claim follows from Corollary 14.11. $\quad\square$

15   Links

A *link* $L \subset S^3$ is a finite set of disjoint circles embedded in S^3. These circles are called the *components* of L. If L is connected, it is called a *knot*. Two links L_1 and L_2 are said to be *ambiently isotopic* if there is an orientation preserving homeomorphism of S^3 onto itself mapping L_1 onto L_2. (It is known that every orientation preserving self-homeomorphism of S^3 is isotopic to the identity, see [13].) A link is *polygonal* if it is the union of a finite number of closed straight-line segments. A link is *tame* if it is ambiently isotopic to a polygonal link. We only consider tame links. If a link parameterized by arc length is continuously differentiable, then it is tame, see [8, Appendix I]. A link $L \subset \mathbb{R}^3 \subset S^3$ is usually specified by a regular projection to a plane in $\mathbb{R}^3$, see [3, 8]. A link is *trivial* if it is ambiently isotopic to a link presented by a diagram without crossings. A trivial 1-component link is called an *unknot*.

Given a link $L \subset \mathbb{R}^3$, let L^* be the image of L under a reflection in a plane. The link L is *amphicheiral* if it is ambiently isotopic to its mirror L^*. A link L is *oriented* if all its components are oriented. An oriented link L is *invertible* if it is ambiently isotopic to the link $-L$ obtained from L by reversing its orientation.

The unknot.   The left-handed trefoil knot.   The figure-eight knot.

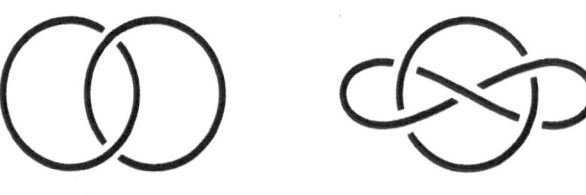

The Hopf link.   The Whitehead link.

Figure 15.1

Exercise 15.1 Show that the figure-eight knot is amphicheiral [8, p. 10]. Convince yourself by experimenting with a piece of rope that the trefoil knot is not amphicheiral. Show that both the left-handed trefoil knot and the figure-eight knot are invertible [8, p. 11]. ◇

We orient S^3 once and forever. A link in S^3 is *ordered* if its components are numbered. Let $L = L_1 \cup \cdots \cup L_l \subset S^3$ be an ordered oriented link. Let $U = \coprod_{i=1}^{l} U_i$ be a regular neighbourhood of L and let $E = S^3 \setminus \text{Int}(U)$ be the *link exterior*. Since E is a deformation retract of $S^3 \setminus L$, the group $H_1(E) = H_1(S^3 \setminus L)$ does not depend on the choice of U. We claim that this group is canonically isomorphic to a free abelian group on l generators. Indeed, the exact homology sequence of the pair (S^3, E) shows that

$$H_1(E) = H_2(S^3, E) = \bigoplus_{i=1}^{l} H_2(U_i, \partial U_i).$$

The regular neighbourhood U_i of L_i can be identified with a solid torus $S^1 \times D^2$ where D^2 is a closed 2-disc. Under the identification $U_i = S^1 \times D^2$ the knot $L_i \subset U_i$ corresponds to the core $S^1 \times \{pt\}$ of $S^1 \times D^2$ where $pt \in \text{Int } D^2$. For any $x \in S^1$, the 2-disc $x \times D^2 \subset U_i$ meets L_i transversally in one point. We orient D^2 so that the sign of this intersection is $+$. The disc $x \times D^2$ with this orientation is called a *meridional disc* in U_i. The oriented loop $x \times \partial D^2 \subset \partial U_i$ is called a *meridian* of L_i. It is easy to see that the isotopy class of the meridian on ∂U_i neither depends on the choice of $x \in S^1$ nor on the choice of identification $U_i = S^1 \times D^2$.

The exact homology sequence of the pair $(U_i, \partial U_i)$ shows that $H_2(U_i, \partial U_i) = \mathbb{Z}$ with generator represented by a meridional disc of L_i. Therefore $H_1(E) = \mathbb{Z}t_1 \oplus \cdots \oplus \mathbb{Z}t_l$ where $t_i \in H_1(E)$ is the homology class of a meridian of L_i. The group ring $\mathbb{Z}[H_1(E)] = \mathbb{Z}[t_1^{\pm 1}, \ldots, t_l^{\pm 1}]$ is the ring of Laurent polynomials in $t_1, \ldots, t_l$.

The group $\pi_1(E) = \pi_1(S^3 \setminus L)$ is called *the group of L*.

Definition 15.2 The *Alexander polynomial of L* is

$$\Delta_L = \Delta_{\pi_1(E)} \in \mathbb{Z}[t_1^{\pm 1}, \ldots, t_l^{\pm 1}] / \pm \{t_1^{k_1} \cdots t_l^{k_l}\}_{k_1, \ldots, k_l \in \mathbb{Z}}.$$

The Alexander polynomial plays a fundamental role in knot theory, see [3, 8, 29].

Applying Corollary 11.9 to the link exterior we obtain the following result.

Corollary 15.3 [23] *Let $L = L_1 \cup \cdots \cup L_l$ be an ordered oriented link with exterior E. If $l = 1$, set $t = t_1 \in H_1(E)$. Then*

$$\Delta_L = \begin{cases} \tau_\mu(E)(t-1) & \text{if} \quad l = 1, \\ \tau_\mu(E) & \text{if} \quad l \geq 2. \end{cases}$$

Corollary 15.3 allows to apply the theory of torsions to the study of the Alexander polynomial of links. E.g., combining this result with the duality for torsions we obtain $\overline{\Delta_L} = \Delta_L$ (Corollary 14.7), a fact known to Seifert [31]. By the systematic use of torsions, many classical theorems on the Alexander polynomial of links can be proved in a unified way and generalized to higher dimensions and to links in more general 3-manifolds. We refer to [34] for details.

The next section and Section 19 present two methods of computing Δ_L.

16 The Fox Differential Calculus

The Fox differential calculus computes Δ_L from a presentation of the group $\pi_1(S^3 \setminus L)$. We start with describing a classical algorithm for writing down a presentation of this group.

16.1 The Wirtinger presentation

Let $L \subset \mathbb{R}^3$ be a link. We may assume that $p = (0, 0, 1) \notin L$. Take p as the base point of $\pi = \pi_1(S^3 \setminus L) = \pi_1(\mathbb{R}^3 \setminus L)$. Fix an orientation of L and a generic projection of L to, say, the x-y plane. The picture of the projection decomposes into a finite union of disjoint oriented embedded arcs $\alpha_1, \ldots, \alpha_n$ (cf. Figure 16.1).

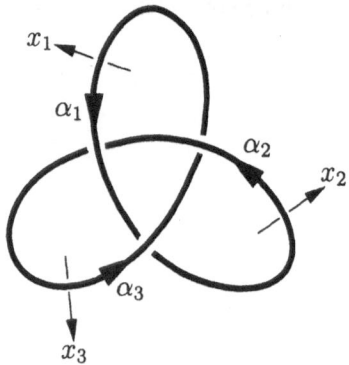

Figure 16.1: The generators of a Wirtinger presentation.

Let us draw a small arrow x_i lying under α_i and crossing α_i in one point from left to right. This arrow determines an element $\xi_i \in \pi$ represented by an oriented triangle whose vertices are p and the endpoints of x_i. Represent an element of π by an elastic string. Pulling all points of this string simultaneously towards p, we see that $\xi_1, \ldots, \xi_n$ generate π. Moreover, at each crossing there is a certain relation among the x_i's which obviously must hold. The two possibilities are depicted in Figure 16.2.

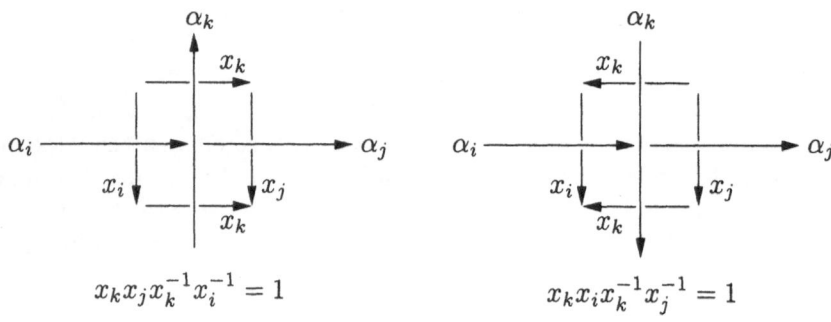

Figure 16.2

Let r_i be the relation holding at the passage from α_i to the next arc α_j.

Theorem 16.1 *The group* $\pi_1(S^3 \setminus L)$ *has a presentation*

$$\pi_1(S^3 \setminus L) = \langle x_1, \ldots, x_n \,|\, r_1, \ldots, r_n \rangle.$$

Moreover, any of the relations r_i *follows from* $\{r_j\}_{j \neq i}$.

The only difficulty in the proof of this theorem is to show that $r_1, \ldots, r_n$ is a complete set of relations in π. For a proof, we refer to [8, Chapter VI] or to [3, Chapter III B]. Theorem 16.1 offers another proof of the fact that for a link L with l components, $H_1(S^3 \setminus L) = \mathbb{Z}^l$. Indeed, let us write $x_i \sim x_j$ if x_i and x_j belong to the same component of L. Abelianizing

$$\pi_1(S^3 \setminus L) = \langle x_1, \ldots, x_n \mid r_1, \ldots, r_n \rangle.$$

we obtain

$$H_1(S^3 \setminus L) = \langle x_1, \ldots, x_n \mid x_i = x_j \text{ if and only if } x_i \sim x_j \rangle = \mathbb{Z}^l.$$

16.2   The Fox differential calculus

Consider the free group on n generators $F = \langle x_1, \ldots, x_n \rangle$. Let

$$\text{aug} \colon \mathbb{Z}[F] \to \mathbb{Z}, \quad \sum_{f \in F} n_f f \mapsto \sum_{f \in F} n_f$$

be the augmentation. An additive homomorphism $\mathcal{D} \colon \mathbb{Z}[F] \to \mathbb{Z}[F]$ is called a *derivative* in $\mathbb{Z}[F]$ if

$$\mathcal{D}(ab) = \mathcal{D}(a) \operatorname{aug}(b) + a\mathcal{D}(b) \quad \text{for all } a, b \in \mathbb{Z}[F]. \tag{16.1}$$

Lemma 16.2 *Let $\mathcal{D}$ be a derivative in $\mathbb{Z}[F]$. Then $\mathcal{D}(k) = 0$ for any integer k, and $\mathcal{D}(a^{-1}) = -a^{-1}\mathcal{D}(a)$ for any $a \in F$.*

Proof. We have that $\mathcal{D}(1) = \mathcal{D}(1 \cdot 1) = \mathcal{D}(1)1 + 1\mathcal{D}(1) = \mathcal{D}(1) + \mathcal{D}(1)$, whence $\mathcal{D}(1) = 0$ and $\mathcal{D}(k) = k\mathcal{D}(1) = 0$. Moreover, $0 = \mathcal{D}(1) = \mathcal{D}(a^{-1}a) = \mathcal{D}(a^{-1}) + a^{-1}\mathcal{D}(a)$, which proves the second claim. □

Any derivative $\mathcal{D}$ in $\mathbb{Z}[F]$ is uniquely determined by its values on the generators of F. Indeed, if $\mathcal{D}$ is given on $x_1, \ldots, x_n$, then, by Lemma 16.2, it is also given on $x_1^{-1}, \ldots, x_n^{-1}$. By (16.1), $\mathcal{D}$ is then known on all products, and hence, by additivity, on all of $\mathbb{Z}[F]$. For the existence part in the following theorem, we refer to [8, Chapter VII].

Theorem 16.3 *For all $j \in \{1, \ldots, n\}$, there exists a unique derivative $\frac{\partial}{\partial x_j}$ in $\mathbb{Z}[F]$ satisfying*

$$\frac{\partial x_i}{\partial x_j} = \delta_{ij} \quad \textit{(Kronecker delta).}$$

Let $\pi = \langle x_1, \ldots, x_n \,|\, r_1, \ldots, r_m \rangle$ be a finitely presented group. Here, $\{x_1, \ldots, x_n\}$ is a set of generators of π, and $\{r_1, \ldots, r_m\}$ is a complete set of relations in π. Thus, r_i are words in the free group $F = \langle x_1, \ldots, x_n \rangle$, and if N is the normal subgroup in F generated by $r_1, \ldots, r_m$, then $\pi = F/N$. Set $H = H_1(\pi) = \pi/[\pi, \pi]$. The projections $F = \langle x_1, \ldots, x_n \rangle \to \pi$ and $\pi \to H$ extend to ring homomorphisms $pr \colon \mathbb{Z}[F] \to \mathbb{Z}[\pi]$ and $h \colon \mathbb{Z}[\pi] \to \mathbb{Z}[H]$, respectively. Set $\rho = h \circ pr \colon \mathbb{Z}[F] \to \mathbb{Z}[H]$.

Definition 16.4 The *Alexander matrix* of the presentation

$$\pi = \langle x_1, \ldots, x_n \,|\, r_1, \ldots, r_m \rangle$$

is the matrix

$$A = \left[\rho \left(\frac{\partial r_i}{\partial x_j} \right) \right]_{\substack{i=1,\ldots,m \\ j=1,\ldots,n}} \quad \text{over } \mathbb{Z}[H].$$

For any $k \geq 0$ set $E_k(\pi) = E_k(A) \subset \mathbb{Z}[H]$ (cf. Section 4.1). It is known that the ideals $E_0(\pi) \subset E_1(\pi) \subset \cdots \subset \mathbb{Z}[H]$ do not depend on the choice of a presentation of π (see, for instance, [8, Chapter VII]). They are called *elementary ideals* of π.

Theorem 16.5 *Let π be a finitely presented group. Let $H = H_1(\pi)$, $G = H/\operatorname{Tors} H$ and $g \colon \mathbb{Z}[H] \to \mathbb{Z}[G]$ be the ring homomorphism induced by the projection $H \to G$. Then*

$$\Delta_\pi = \gcd(g(E_1(\pi))) \in \mathbb{Z}[G]/\pm G. \tag{16.2}$$

Proof. Let π be presented by $\langle x_1, \ldots, x_n \,|\, r_1, \ldots, r_m \rangle$. Associate to this presentation a 2-dimensional CW-complex X as follows. The 0-skeleton X^0 consists of one point e^0. The 1-skeleton $X^1 = \bigvee_{i=1}^n S_i^1$ is a wedge of n oriented circles corresponding to the generators x_i (cf. Figure 16.3).

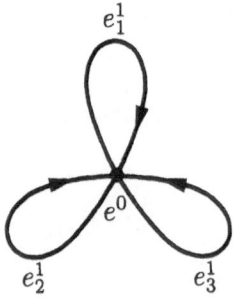

Figure 16.3: A wedge of 3 circles.

It is well known that $\pi_1(X_1, e^0) = \langle x_1, \ldots, x_n \rangle$. Each word r_j determines a loop γ_j in X_1. The CW-complex X is obtained from X_1 by gluing m closed 2-cells $\bar{e}_j^2$ along the loops γ_j. Consider the natural homomorphism

$$\varphi\colon \langle x_1, \ldots, x_n \rangle \to \pi_1(X), \qquad x_i \mapsto [\bar{e}_i^1].$$

Clearly, φ is surjective and the words r_j are contained in its kernel. Hence, φ induces an epimorphism $\pi \to \pi_1(X)$. By the Seifert–van Kampen theorem, it is an isomorphism, see [21, Chapter VII, Theorem 2.1].

Consider the universal covering $p\colon \tilde{X} \to X$ of X. Lift the oriented cells e^0, e_i^1, e_j^2 to cells $\tilde{e}^0, \tilde{e}_i^1, \tilde{e}_j^2$ in $\tilde{X}$, $i = 1, \ldots, n$, $j = 1, \ldots, m$. We assume that the oriented arc $\tilde{e}_i^1$ starts at $\tilde{e}^0$, $i = 1, \ldots, n$, and that the lift $\tilde{f}_j$ of the characteristic map $f_j\colon D^2 \to \bar{e}_j^2$ maps the point $(1, 0) \in D^2$ to $\tilde{e}^0$, $j = 1, \ldots, m$. We have

$$C(\tilde{X}) = \left(0 \to \bigoplus_{j=1}^{m} \mathbb{Z}[\pi]\, \tilde{e}_j^2 \xrightarrow{\partial_1} \bigoplus_{i=1}^{n} \mathbb{Z}[\pi]\, \tilde{e}_i^1 \to \mathbb{Z}[\pi]\, \tilde{e}^0 \to 0 \right).$$

Assume that ∂_1 is given by the matrix (α_{ji}):

$$\partial_1 \tilde{e}_j^2 = \sum_{i=1}^{n} \alpha_{ji} \tilde{e}_i^1, \qquad \alpha_{ji} \in \mathbb{Z}[\pi].$$

Claim 16.6 *We have* $\alpha_{ji} = pr\left(\dfrac{\partial r_j}{\partial x_i} \right) \in \mathbb{Z}[\pi]$.

Proof. Fix j and write

$$r_j = x_{j_1}^{\varepsilon_1} \cdots x_{j_k}^{\varepsilon_k}, \qquad \varepsilon_i = \pm 1. \tag{16.3}$$

Recall from Section 5.6 that we may compute $\partial_1 \tilde{e}_j^2$ by walking along the boundary of $\tilde{e}_j^2$ and collecting the contributions of the boundary segments. By (16.3) the first boundary segment s_1 lies over $e_{j_1}^1$. If $\varepsilon_1 = +1$, then $s_1 = \tilde{e}_{j_1}^1$. If $\varepsilon_1 = -1$, then $s_1 = pr(x_{j_1}^{-1})\, \tilde{e}_{j_1}^1$. In the first case, the contribution of s_1 to $\partial_1 \tilde{e}_j^2$ is $+\tilde{e}_{j_1}^1$, and in the second case it is $-pr(x_{j_1}^{-1})\, \tilde{e}_{j_1}^1$. Assume that we already walked along the first $l-1$ segments of the boundary of $\tilde{e}_j^2$. The segment s_l lies over $e_{j_l}^1$. If $\varepsilon_l = +1$, then $s_l = pr(x_{j_1}^{\varepsilon_1} \cdots x_{j_{l-1}}^{\varepsilon_{l-1}})\, \tilde{e}_{j_l}^1$. If $\varepsilon_l = -1$, then $s_l = pr(x_{j_1}^{\varepsilon_1} \cdots x_{j_{l-1}}^{\varepsilon_{l-1}})\, pr(x_{j_l}^{-1})\, \tilde{e}_{j_l}^1$. In the first case, the contribution of s_l to $\partial_1 \tilde{e}_j^2$ is $pr(x_{j_1}^{\varepsilon_1} \cdots x_{j_{l-1}}^{\varepsilon_{l-1}})\, \tilde{e}_{j_l}^1$, and in the second case it is $pr(x_{j_1}^{\varepsilon_1} \cdots x_{j_{l-1}}^{\varepsilon_{l-1}}(-x_{j_l}^{-1}))\, \tilde{e}_{j_l}^1$. Observe now that by (16.1),

$$\frac{\partial r_j}{\partial x_i} = \frac{\partial}{\partial x_i} x_{j_1}^{\varepsilon_1} + x_{j_1}^{\varepsilon_1} \frac{\partial}{\partial x_i} x_{j_2}^{\varepsilon_2} + \cdots + x_{j_1}^{\varepsilon_1} \cdots x_{j_{k-1}}^{\varepsilon_{k-1}} \frac{\partial}{\partial x_i} x_{j_k}^{\varepsilon_k}.$$

Moreover, $\frac{\partial}{\partial x_i} x_j = \delta_{ij}$ and $\frac{\partial}{\partial x_i} x_j^{-1} = \delta_{ij}(-x_j^{-1})$ for all j. The claim thus follows.   □

Recall from Section 11 that

$$\Delta_\pi = \operatorname{ord} A_1(X) = \operatorname{ord} H_1(\bar{X}) = \Delta_0(H_1(\bar{X})) \qquad (16.4)$$

where $\bar{X}$ is the maximal free abelian covering of X. Let $\bar{X}^0 \subset \bar{X}$ be the preimage of $e^0 \in X$. The sequence

$$C_2(\bar{X}) \xrightarrow{\partial_1} C_1(\bar{X}) \to H_1(\bar{X}, \bar{X}^0) \to 0$$

is exact and yields a presentation of the $\mathbb{Z}[G]$-module $H_1(\bar{X}, \bar{X}^0)$. Therefore, the matrix $[gh(\alpha_{ji})]_{i,j}$ is a presentation matrix of $H_1(\bar{X}, \bar{X}^0)$. Claim 16.6 implies that

$$[gh(\alpha_{ji})] = \left[gh\, pr\left(\frac{\partial r_j}{\partial x_i}\right)\right] = \left[g\rho\left(\frac{\partial r_j}{\partial x_i}\right)\right] = g(A)$$

where A is the matrix considered in Definition 16.4. Hence,

$$E_k(H_1(\bar{X}, \bar{X}^0)) = g(E_k(\pi)) \quad \text{for all } k$$

and so

$$\Delta_k(H_1(\bar{X}, \bar{X}^0)) = \gcd(E_k(H_1(\bar{X}, \bar{X}^0))) = \gcd(g(E_k(\pi))) \quad \text{for all } k.$$

In particular, $\Delta_1(H_1(\bar{X}, \bar{X}^0)) = \gcd(g(E_1(\pi)))$. Equality (16.2) now follows from (16.4) and the general equality

$$\Delta_{k-1}(H_1(\bar{X})) = \Delta_k(H_1(\bar{X}, \bar{X}^0)), \quad k \geq 1.$$

By Lemma 4.9, it suffices to prove that the $\mathbb{Z}[G]$-modules $H_1(\bar{X})$, $H_1(\bar{X}, \bar{X}^0)$ have isomorphic $\mathbb{Z}[G]$-torsion submodules and $\operatorname{rk} H_1(\bar{X}, \bar{X}^0) = \operatorname{rk} H_1(\bar{X}) + 1$. The first property follows from the exact homology sequence

$$0 \to H_1(\bar{X}) \to H_1(\bar{X}, \bar{X}^0) \to H_0(\bar{X}^0) \to H_0(\bar{X}) \to 0$$

using the obvious fact that $H_0(\bar{X}^0) = \mathbb{Z}[G]$. To compute the rank, one tensorizes with $Q(G)$ which kills $H_0(\bar{X}) = \mathbb{Z}$ and makes the sequence split. Hence, $\operatorname{rk} H_1(\bar{X}, \bar{X}^0) = \operatorname{rk} H_1(\bar{X}) + 1$.   □

The following observations are useful in computing the Alexander polynomial of links via the Fox calculus.

Let $\pi = \langle x_1, \ldots, x_n \,|\, r_1, \ldots, r_n \rangle$ be the Wirtinger presentation of the group of a link L with l components. As we have seen after Theorem 16.1, the abelianization $\pi \to H_1(\pi) = \mathbb{Z}t_1 \oplus \cdots \oplus \mathbb{Z}t_l$ maps all generators x_i belonging to the

k'th component to t_k. In particular, if $\rho: \mathbb{Z}[\langle x_1, \ldots, x_n \rangle] \to \mathbb{Z}[H_1(\pi)]$ is the natural projection, then $\rho(x_i) = \rho(x_j)$ whenever x_i and x_j belong to the same component of L.

Let r and s be words in $\langle x_1, \ldots, x_n \rangle$ and assume that the relation $r = s$ holds in π. By Lemma 16.2,

$$\frac{\partial (rs^{-1})}{\partial x_j} = \frac{\partial r}{\partial x_j} + r\frac{\partial s^{-1}}{\partial x_j} = \frac{\partial r}{\partial x_j} - rs^{-1}\frac{\partial s}{\partial x_j}.$$

Since $\rho(rs^{-1}) = 1$, we conclude that

$$\rho\left(\frac{\partial rs^{-1}}{\partial x_j}\right) = \rho\left(\frac{\partial r}{\partial x_j} - \frac{\partial s}{\partial x_j}\right). \tag{16.5}$$

Notation 16.7 Let $p, q \in \mathbb{Q}(t_1, \ldots, t_l)$. We write

$$p(t_1, \ldots, t_l) \doteq q(t_1, \ldots, t_l)$$

if $p = \pm t_1^{\nu_1} \cdots t_l^{\nu_l} q$ with $\nu_1, \ldots, \nu_l \in \mathbb{Z}$.

Examples 16.8
  1. Let K be an unknot. We present K by the projection in Figure 16.4.

Figure 16.4: The unknot.

The only relation $x_1 x_1 x_1^{-1} x_1^{-1} = 1$ is trivial. Hence,

$$\pi = \pi_1(S^3 \setminus K) \cong \langle x_1 \rangle = \mathbb{Z}.$$

The Alexander matrix is $A = [0]$ and so $E_1(\pi) = \mathbb{Z}[t^{\pm 1}]$ and $\Delta_K = \Delta_\pi \doteq 1 \in \mathbb{Z}[t^{\pm 1}]$.

  2. Let K be the (left-handed) trefoil knot (cf. Figure 16.1). The set of relations is
$$\{x_1 x_3^{-1} x_1^{-1} x_2, \ x_2 x_1^{-1} x_2^{-1} x_3, \ x_3 x_2^{-1} x_3^{-1} x_1\}.$$

Observe that each relation is indeed a consequence of the other two. The Wirtinger presentation of $\pi = \pi_1(S^3 \setminus K)$ may be simplified as follows.

$$
\begin{aligned}
\pi &= \langle x_1, x_2, x_3 \mid x_2 = x_1 x_3 x_1^{-1}, \ x_3 = x_2 x_1 x_2^{-1} \rangle \\
&= \langle x_1, x_2 \mid x_2 = x_1 x_2 x_1 x_2^{-1} x_1^{-1} \rangle \\
&= \langle x, y \mid xyx = yxy \rangle.
\end{aligned}
$$

By (16.5), the Alexander matrix $A = (a_1, a_2)$ of this presentation is given by

$$
a_1 = \rho\left(\frac{\partial(xyx)}{\partial x} - \frac{\partial(yxy)}{\partial x} \right) = \rho(1 + xy - y) = t^2 - t + 1,
$$

$$
a_2 = \rho\left(\frac{\partial(xyx)}{\partial y} - \frac{\partial(yxy)}{\partial y} \right) = \rho(x - 1 - yx) = -t^2 + t - 1.
$$

Hence, $E_1(\pi) = \langle t^2 - t + 1 \rangle$ and $\Delta_K \doteq t^2 - t + 1$.

3. Let K be the figure-eight knot. A presentation of $\pi = \pi_1(S^3 \setminus K)$ is given by generators x, y, w, z and relations $xw^{-1}x^{-1}y, \ wx^{-1}w^{-1}z, \ zxz^{-1}y$ (cf. Figure 16.5).

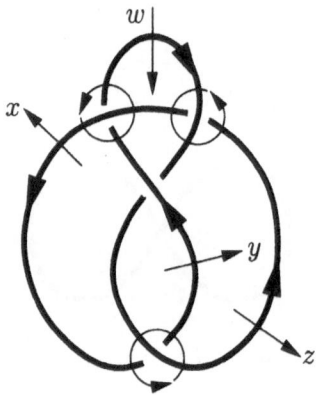

Figure 16.5

Hence,

$$
\begin{aligned}
\pi &= \langle x, y, w, z \mid w = x^{-1}yx, \ z = wxw^{-1}, \ y = zxz^{-1} \rangle \\
&= \langle x, y, z \mid z = x^{-1}yxy^{-1}x, \ y = zxz^{-1} \rangle \\
&= \langle x, y \mid y = x^{-1}yxy^{-1}xyx^{-1}y^{-1}x \rangle \\
&= \langle x, y \mid yx^{-1}yxy^{-1} = x^{-1}yxy^{-1}x \rangle.
\end{aligned}
$$

We compute the Alexander matrix $A = (a_1, a_2)$:

$$a_1 = \rho \frac{\partial}{\partial x} \left(yx^{-1}yxy^{-1} - x^{-1}yxy^{-1}x \right) = t - 3 + t^{-1},$$

$$a_2 = \rho \frac{\partial}{\partial y} \left(yx^{-1}yxy^{-1} - x^{-1}yxy^{-1}x \right) = -t + 3 - t^{-1}.$$

Hence, $E_1(\pi) = \langle t^2 - 3t + 1 \rangle$ and $\Delta_K \doteq t^2 - 3t + 1$.

These computations show that the unknot, the trefoil knot and the figure-eight knot are mutually not ambiently isotopic.

4. Let L be a trivial link with l components, $l \geq 2$. Then $\pi = \pi_1(S^3 \setminus L)$ is the free group of rank l. Hence, $E_1(\pi) = 0$ and $\Delta_L \doteq 0$.

5. Let L be the Hopf link. Then $\pi = \pi_1(S^3 \setminus L) = \langle x, y \mid xy = yx \rangle = \mathbb{Z}^2$. Let $A = (a_1, a_2)$ and $\rho(x) = t_1$, $\rho(y) = t_2$. Then,

$$a_1 = \rho \frac{\partial}{\partial x} (xy - yx) = 1 - t_2, \qquad a_2 = \rho \frac{\partial}{\partial y} (xy - yx) = 1 - t_1.$$

Hence, $E_1(\pi) = \langle t_1 - 1, t_2 - 1 \rangle$ and $\Delta_L \doteq 1$.

6. Let L be the Whitehead link drawn in Figure 16.6.

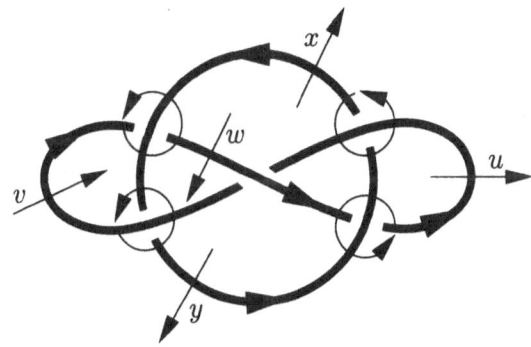

Figure 16.6

Then

$$
\begin{aligned}
\pi_1(S^3 \setminus L) &= \langle u, v, w, x, y \mid ux^{-1}u^{-1}y,\ vxv^{-1}y^{-1},\ xvx^{-1}w^{-1},\ yu^{-1}y^{-1}w \rangle \\
&= \langle u, v, x, y \mid ux^{-1}u^{-1}y,\ vxv^{-1}y^{-1},\ xvx^{-1}yu^{-1}y^{-1} \rangle \\
&= \langle u, x, y \mid ux^{-1}u^{-1}y,\ x^{-1}yuy^{-1}xyu^{-1}y^{-1}xy^{-1} \rangle \\
&= \langle u, y \mid uyuy^{-1}u^{-1}yuy = yuyu^{-1}y^{-1}uyu \rangle.
\end{aligned}
$$

Let $A = (a_1, a_2)$ and $\rho(u) = t_1$, $\rho(y) = t_2$. Then

$$
\begin{aligned}
a_1 &= \rho\frac{\partial}{\partial u}\left(uyuy^{-1}u^{-1}yuy - yuyu^{-1}y^{-1}uyu\right) \\
&= 1 - t_1 - 2t_2 + 2t_1t_2 + t_2^2 - t_1t_2^2 = (1-t_1)(1-t_2)^2, \\
a_2 &= \rho\frac{\partial}{\partial y}\left(uyuy^{-1}u^{-1}yuy - yuyu^{-1}y^{-1}uyu\right) \\
&= -1 + t_2 + 2t_1 - 2t_1t_2 - t_1^2 + t_1^2t_2 = -(1-t_2)(1-t_1)^2.
\end{aligned}
$$

Hence, $\Delta_L \doteq (1-t_1)(1-t_2)$ and $E_1(\pi_1(S^3 \setminus L)) = \langle \Delta_L(1-t_1), \Delta_L(1-t_2)\rangle$.

These computations show that the trivial link with 2 components, the Hopf link and the Whitehead link are mutually not ambiently isotopic.

For other applications of the Fox differential calculus, see [1, Chapter 3].

17 Computing $\tau(M^3)$ from the Alexander polynomial of links

17.1 Surgery on links

We first recall the classical notion of a linking number of knots in S^3. If K, K' are disjoint oriented knots in S^3, then K' represents an element $[K'] \in H_1(S^3 \backslash K)$. The group $H_1(S^3 \backslash K) = \mathbb{Z}$ is generated by the homology class t of the meridian of K. Thus, (in multiplicative notation) $[K'] = t^n$ with $n \in \mathbb{Z}$. The number n is denoted by $\mathrm{lk}(K, K')$ and called the *linking number* of K and K'. For instance, the linking number of an oriented knot and its meridian is equal to $+1$. Clearly, $\mathrm{lk}(-K, K') = \mathrm{lk}(K, -K') = -\mathrm{lk}(K, K')$ and $\mathrm{lk}(-K, -K') = \mathrm{lk}(K, K')$. The linking number is symmetric: $\mathrm{lk}(K, K') = \mathrm{lk}(K', K)$.

A link $L = L_1 \cup \cdots \cup L_l \subset S^3$ is said to be *framed* if each component L_i is endowed with an integer k_i called *the framing* of L_i. Given a framed l-component link L as above we define a closed 3-manifold as follows. Let $U = \coprod_{i=1}^{l} U_i$ be a regular neighbourhood of L. The neighbourhood U_i of L_i is homeomorphic to $S^1 \times D^2$ where D^2 is the unit disc in $\mathbb{C}$ and $S^1 = \partial D^2$. Consider a homeomorphism $f_i : S^1 \times D^2 \to U_i$ such that for one (and then for any) point $pt \in \partial D^2$ the linking number of the knots $f_i(S^1 \times pt)$ and $L_i = f_i(S^1 \times 0)$ is equal to k_i. Here, 0 is the center of D^2 and the linking number is computed with respect to the orientations of $f_i(S^1 \times pt)$, $f_i(S^1 \times 0)$ induced by one and the same orientation of S^1. Such a homeomorphism f_i can be obtained from an arbitrary homeomorphism $S^1 \times D^2 \to U_i$ by composing it with a homeomorphism $U_i \to U_i$ rotating the meridional discs around L_i. The exterior $E = S^3 \backslash \mathrm{Int}(U)$ of L is a compact 3-manifold with boundary $\partial E = \coprod_{i=1}^{l} \partial U_i$. Let us glue a copy of $D^2 \times S^1$ to each component ∂U_i of ∂E along the homeomorphism $f_i|_{S^1 \times S^1} : \partial(D^2 \times S^1) \to \partial U_i$. This yields a closed

3-manifold M containing E. We say that M is obtained from S^3 by *surgery along* L. The manifold M is connected and orientable.

The equality $H_1(M, E) = \oplus_{i=1}^l H_1(U_i, \partial U_i) = 0$ implies that the inclusion homomorphism $H_1(E) \rightarrow H_1(M)$ is surjective. Orient $L = L_1 \cup \cdots \cup L_l$ in an arbitrary way. Then the meridians of $L_1, \ldots, L_l$ lie in $E \subset M$ and represent certain generators $h_1, \ldots, h_l \in H_1(M)$ of $H_1(M)$. They satisfy l generating relations (in multiplicative notation)

$$\prod_{j=1}^l h_j^{k_{i,j}} = 1$$

where i runs over $\{1, \ldots, l\}$ and

$$k_{i,j} = \begin{cases} \mathrm{lk}(L_i, L_j) & \text{if} \quad i \neq j, \\ k_i & \text{if} \quad i = j. \end{cases}$$

In particular, $b_1(M) = \mathrm{rk}\, H_1(M) \neq 0$ if and only if $\det(k_{i,j}) = 0$.

A fundamental theorem due to W. B. R. Lickorish and A. H. Wallace says that every closed connected orientable 3-manifold M can be obtained by surgery along a framed link $L \subset S^3$ (see [29, Chapter 9] for a proof). Using so-called handle slidings one can show that if $b_1(M) \geq 1$ then we can choose $L = L_1 \cup \cdots \cup L_l \subset S^3$ so that the classes $h_1, \ldots, h_l \in H_1(M)$ represented by the meridians of $L_1, \ldots, L_l$ all have infinite order in $H_1(M)$ (see [37, p. 594]).

17.2 Computing $\tau(M^3)$ **from the Alexander polynomial**

Lemma 17.1 *Let H be a finitely generated abelian group and $h \in H$ be an element of infinite order. Then $h - 1$ is invertible in $Q(H) = Q(\mathbb{Z}[H])$.*

Proof. Recall from Section 13 that $Q(H) = \oplus_{i=1}^k Q(R_i)$ is a direct sum of fields. Let $p_i \colon Q(H) \rightarrow Q(R_i)$ be the projection. It follows from the construction of the splitting $Q(H) = \oplus_{i=1}^k Q(R_i)$ that $p_i(h) \neq 1 \in Q(R_i)$, $i = 1, \ldots, k$. Therefore, $h - 1$ is invertible in $Q(H)$. $\qquad\square$

Theorem 17.2 [33] *Let M be a closed 3-manifold obtained by surgery along a framed oriented link $L = L_1 \cup \cdots \cup L_l \subset S^3$. Let $h_1, \ldots, h_l \in H = H_1(M)$ be the classes represented by the meridians of $L_1, \ldots, L_l$. If $b_1(M) \geq 1$ and $h_1, \ldots, h_l$ all have infinite order in H, then we have the following equality in $Q(H)/\pm H$:*

$$\tau(M) = \begin{cases} \Delta_L(h_1)\,(h_1 - 1)^{-2} & \text{if} \quad l = 1, \\ \Delta_L(h_1, \ldots, h_l) \prod_{i=1}^l (h_i - 1)^{-1} & \text{if} \quad l \geq 2. \end{cases}$$

This theorem follows from the interpretation of the Alexander polynomial as the torsion of E and the multiplicativity of torsions applied to the pair (M, E) where E is the exterior of L in S^3. The key argument is given by the following lemma.

Lemma 17.3 *Let* $U = S^1 \times D^2$ *be a solid torus. Set* $G = H_1(U)$. *Let* $\mathbb{F}$ *be a field and* $\varphi \colon \mathbb{Z}[G] \to \mathbb{F}$ *be a ring homomorphism such that* $\varphi(G) \neq 1$. *Then* $H^\varphi_*(U, \partial U) = 0$ *and* $\tau_\varphi(U, \partial U) = (\varphi(h) - 1)^{-1} \in \mathbb{F}^*/\pm\varphi(G)$ *where* h *is a generator of the infinite cyclic group* G.

Proof. We choose a CW-decomposition of S^1 consisting of one 0-cell and one 1-cell. Choose a CW-decomposition of D^2 consisting of one 0-cell, one 1-cell and one 2-cell. Provide $U = S^1 \times D^2$ with the product CW-decomposition. Note that U is obtained from ∂U by adjoining one 2-cell and one 3-cell. The relative cellular chain complex $C^\varphi(U, \partial U) = (0 \to C_3 \to C_2 \to 0)$ has only one non-zero boundary homomorphism $C_3 \to C_2$ given (in appropriate bases) by the (1×1)-matrix $(\varphi(h) - 1)$. This implies both claims of the lemma. $\square$

Theorem 17.2 allows to compute the maximal abelian torsion of a closed connected oriented 3-manifold M with $b_1(M) \geq 1$. Without the assumptions on $b_1(M)$ and $h_1, \ldots, h_l$ in Theorem 17.2 we can obtain the following weaker result. Let $H = H_1(M)$ and

$$I = I_L = \prod_{i=1}^{l} (h_i - 1) \, Q(H) \subset Q(H)$$

where $h_i \in H$ is represented by the meridian of L_i. It is clear that I is an ideal of $Q(H)$. If $Q(H) = \oplus_{j \in J} F_j$ is a splitting of $Q(H)$ into a direct sum of fields, then $I = \oplus_{j \in J_0} F_j$ where $J_0 \subset J$ consists of those $j \in J$ for which the projections of $h_1, \ldots, h_l$ to F_j are all distinct from 1. Let $p \colon Q(H) \to I$ be the projection with kernel $\oplus_{j \in J \setminus J_0} F_j$. Then $p(h_i - 1)$ is invertible in the ring I for all i and

$$p(\tau(M)) = \begin{cases} p(\Delta_L(h_1)) \, (p(h_1 - 1))^{-2} & \text{if} \quad l = 1, \\ p(\Delta_L(h_1, \ldots, h_l)) \prod_{i=1}^{l} (p(h_i - 1))^{-1} & \text{if} \quad l \geq 2. \end{cases}$$

Under the assumptions stated in Theorem 17.2, $I = Q(H)$ so that p is the identity homomorphism. There is one more case where this formula is strong enough to compute $\tau(M)$. Namely, assume that $b_1(M) = 0$, the group $H = H_1(M)$ is cyclic and $h_i \in H$ is a generator of H for all $i = 1, \ldots, l$. By Corollary 12.7 and Remark 12.8, $I_L = \text{Ker}(\text{aug} \colon \mathbb{Q}[H] \to \mathbb{Q}) = I(H)$ is the augmentation ideal of $\mathbb{Q}[H]$. Moreover, we have seen in Section 13 that $\tau(M)$ lies in $I(H)$. So $p(\tau(M)) = \tau(M)$ and the formula above computes $\tau(M)$ from Δ_L.

Examples and Remarks 17.4

1. Let M be obtained by surgery on S^3 along an oriented knot $K \subset S^3$ with framing 0. Then $H_1(M)$ is an infinite cyclic group $\{h^n\}_{n \in \mathbb{Z}}$ with generator h represented by the meridian of K. By Theorem 17.2,

$$\tau(M) = \Delta_K(h)\,(h-1)^{-2} \in Q(H)/\pm H. \tag{17.1}$$

In this case the group $H_1(M)$ has no torsion so that the maximal abelian torsion coincides with the Milnor torsion.

2. Let M be obtained by surgery on S^3 along an oriented knot $K \subset S^3$ with framing $p \geq 2$ (not necessarily prime). Then $H = H_1(M)$ is a cyclic group $\{h^n\}_{n \in \mathbb{Z}/p\mathbb{Z}}$ whose generator h is represented by the meridian of K. By the discussion above,

$$\tau(M) = \Delta_K(h)\,(h-1)^{-2} \in I(H)/\pm H,$$

while, $H_1(M)$ being finite, the Milnor torsion of M is 0. In the case where K is an unknot, the manifold M is the lens space $L(p;1,1)$ and we obtain the same formula as in Example 13.1.

3. A realization theorem due to Seifert [31] says that a Laurent polynomial $\Delta(t) \in \mathbb{Z}[t, t^{-1}]$ can be realized as the Alexander polynomial of a knot if and only if $\Delta(t^{-1}) \doteq \Delta(t)$ and $\Delta(1) = \pm 1$. Using (17.1) we obtain a vast class of 3-manifolds with non-trivial torsions. An analogous characterization for the Alexander polynomials of links is not known. Note, however, a theorem of Bailey [18, p. 92] which claims that for any Laurent polynomial $\Delta \in \mathbb{Z}[t_1^{\pm 1}, t_2^{\pm 1}]$ such that $\Delta(t_1^{-1}, t_2^{-1}) = t_1^{\nu_1} t_2^{\nu_2} \Delta(t_1, t_2)$ with $\nu_1, \nu_2 \in 2\mathbb{Z}$, the product $(t_1 - 1)(t_2 - 1)\Delta$ can be realized as the Alexander polynomial of a 2-component link $L_1 \cup L_2 \subset S^3$ with $\mathrm{lk}(L_1, L_2) = 0$. If M is the result of surgery on this link with zero framing, then $H_1(M) = \mathbb{Z}^2$ and by Theorem 17.2, $\tau(M) = \tau_\mu(M) = \Delta$.

4. The definition of the Alexander polynomial of a link and Theorem 17.2 directly extend to links in homology 3-spheres.

5. Let M be a compact connected orientable 3-manifold whose boundary is void or consists of tori. Set $H = H_1(M)$. The first elementary ideal $E = E_1(\pi_1(M)) \subset \mathbb{Z}[H]$ of the fundamental group of M can be computed in terms of the torsion $\tau(M)$, see [33, 36]: Let $I = \mathrm{Ker}(\mathrm{aug}\colon \mathbb{Z}[H] \to \mathbb{Z})$ be the augmentation ideal of the ring $\mathbb{Z}[H]$. If $b_1(M) \geq 1$ and $\partial M = \emptyset$ then $E = \tau(M)I^2$. If $b_1(M) \geq 1$ and $\partial M \neq \emptyset$ then $E = \tau(M)I$. If $b_1(M) = 0$ then E is the pre-image of $\tau(M)I^2$ under the homomorphism $\mathbb{Z}[H] \to \mathbb{Q}[H]$ sending $h \in H$ to $h - \frac{1}{|H|}\Sigma$ where $\Sigma = \sum_{f \in H} f \in \mathbb{Z}[H]$. Note that by Lemma 11.12 the assumption $b_1(M) = 0$ implies $\partial M = \emptyset$. A similar result holds in the non-orientable case. As an exercise, the reader may deduce Corollary 11.9 and Theorem 14.12 from this computation of E.

Chapter III
Refined Torsions

18 The sign-refined torsion

Let $C = (0 \to C_m \xrightarrow{\partial_{m-1}} \cdots \xrightarrow{\partial_0} C_0 \to 0)$ be a based chain complex over a field $\mathbb{F}$. Consider the residues $(\mathrm{mod}\, 2)$

$$
\begin{aligned}
\beta_i(C) &= \dim H_i(C) - \dim H_{i-1}(C) + \cdots + (-1)^i \dim H_0(C) \quad (\mathrm{mod}\, 2), \\
\gamma_i(C) &= \dim C_i - \dim C_{i-1} + \cdots + (-1)^i \dim C_0 \quad (\mathrm{mod}\, 2), \\
N(C) &= \sum_{i=0}^{m} \beta_i(C) \gamma_i(C) \in \mathbb{Z}/2\mathbb{Z}.
\end{aligned}
$$

As in Section 3.1, let c_i be a basis of C_i, b_i be a basis of $B_i = \mathrm{Im}\, \partial_i \subset C_i$ and h_i be a basis of $H_i(C)$, $i = 0, \ldots, m$. Set

$$
\check{\tau}(C, \{c_i\}, \{h_i\}) = (-1)^{N(C)} \tau(C) = (-1)^{N(C)} \prod_{i=0}^{m} [b_i h_i b_{i-1}/c_i]^{(-1)^{i+1}} \in \mathbb{F}^*.
$$

Recall that neither $\tau(C)$ nor $\check{\tau}(C)$ depend on the choice of the bases b_i. If C is acyclic, then $\beta_i(C) = 0$ for all i and $\check{\tau}(C) = \tau(C)$.

Let X be a finite connected CW-complex. Consider the finite dimensional real vector space

$$
H_*(X; \mathbb{R}) = \bigoplus_{i \geq 0} H_i(X; \mathbb{R}).
$$

An orientation of $H_*(X; \mathbb{R})$ is called a *homology orientation* of X. We assume from now on that X is homology oriented.

Set $H = H_1(X; \mathbb{Z})$. Let $\varphi \colon \mathbb{Z}[H] \to \mathbb{F}$ be a ring homomorphism and let $\widehat{X}$ be the maximal abelian covering of X. Order and orient the cells of X and denote the resulting family of ordered oriented cells by $\{e\}$. We lift $\{e\}$ to an ordered and oriented family $\{\hat{e}\}$ of cells in $\hat{X}$. This gives us a basis of $C(\hat{X})$ over $\mathbb{Z}[H]$ and hence a basis $\hat{e}$ of $C^{\varphi}(X) = \mathbb{F} \otimes_{\varphi} C(\hat{X})$. We assume that $H_*^{\varphi}(X) = 0$. The family $\{e\}$ determines a basis e of $C = C(X; \mathbb{R})$. Choose a basis h_i in $H_i(X; \mathbb{R})$, $i = 0, 1, \ldots$, such that the sequence $h = (h_0, h_1, \ldots)$ is a positive basis in the oriented vector space $H_*(X; \mathbb{R})$. Then $\check{\tau}(C, e, h) \in \mathbb{R}^* = \mathbb{R} \setminus \{0\}$. Set

$$\tau_{\varphi}^0(X, \hat{e}, h) = \operatorname{sign}\left(\check{\tau}(C, e, h)\right) \cdot \tau_{\varphi}(X, \hat{e}) \in \mathbb{F}^*.$$

Lemma 18.1 $\tau_{\varphi}^0(X, \hat{e}, h)$ *is independent of the order of the cells of X, their orientations, and the choice of h.*

Proof. Observe that $N(C)$ does not depend on any of these choices. Let e_1 and e_2 be consecutive i-cells in $\{e\}$ and let $\{e'\}$ be the family obtained from $\{e\}$ by interchanging e_1 and e_2. By Remark 1.4.1, $\tau(C, e', h) = -\tau(C, e, h)$. Similarly, $\tau_{\varphi}(X, \hat{e}') = -\tau_{\varphi}(X, \hat{e})$, and so $\tau_{\varphi}^0(X, \hat{e}', h) = \tau_{\varphi}^0(X, \hat{e}, h)$. A similar argument shows that $\tau_{\varphi}^0(X, \hat{e}, h)$ does not depend on the orientations of the cells in $\{e\}$. Finally, let $h = (h_0, h_1, \ldots)$ and $h' = (h_0', h_1', \ldots)$ be positive bases in $H_*(X; \mathbb{R})$. Then

$$\tau_{\varphi}^0(X, \hat{e}, h') = \left(\prod_{i \geq 0} \operatorname{sign}[h_i'/h_i]\right) \tau_{\varphi}^0(X, \hat{e}, h).$$

Since both h and h' are positive, $\prod_{i \geq 0} \operatorname{sign}[h_i'/h_i] = +1$. $\square$

Definition 18.2 The *sign-refined torsion* $\tau_{\varphi}^0(X)$ of the homology oriented finite connected CW-complex X is the image of $\tau_{\varphi}^0(X, \hat{e}, h)$ in $\mathbb{F}^*/\varphi(H)$.

It is clear that $\tau_{\varphi}^0(-X) = -\tau_{\varphi}^0(X)$ where $-X$ is X with opposite homology orientation.

Theorem 18.3 $\tau_{\varphi}^0(X)$ *is invariant under simple homotopy equivalences preserving the homology orientation.*

Proof. By Theorem 8.7, a simple homotopy equivalence is a composition of elementary expansions and elementary collapses. Since an elementary collapse is the homotopy inverse of an elementary expansion, the theorem follows from

Lemma 18.4 *Let $i: X \hookrightarrow Y$ be an elementary expansion. Assume that both X and Y are homology oriented and that the inclusion isomorphism $i_*: H_*(X; \mathbb{R}) \to H_*(Y; \mathbb{R})$ is orientation preserving. If $H_*^{\varphi}(X) = H_*^{\varphi}(Y) = 0$, then $\tau_{\varphi}^0(X) = \tau_{\varphi}^0(Y)$.*

Here we denote by the same letter φ a ring homomorphism $\mathbb{Z}[H_1(Y)] \to \mathbb{F}$ and its composition with the inclusion isomorphism $\mathbb{Z}[H_1(X)] \to \mathbb{Z}[H_1(Y)]$.

Proof of Lemma 18.4. The CW-complex Y is obtained from X by adjoining two cells e^k and e^{k+1} as in Section 8. Set $C = C^\varphi(X)$ and $C' = C^\varphi(Y)$. Clearly, C is a subcomplex of C'. Let $\hat{e}^k$ and $\hat{e}^{k+1}$ be elements of C' represented by oriented cells in the maximal abelian covering of Y lying over e^k and e^{k+1}, respectively. We can assume that $\hat{e}^k$ and $\hat{e}^{k+1}$ are incident, and we choose the orientations of e^k, e^{k+1} so that their incidence number (and therefore the incidence number of $\hat{e}^k$, $\hat{e}^{k+1}$) is $+1$. Lift a family of ordered and oriented cells $\{e\}$ in X to bases c_i in $C_i = C_i^\varphi(X)$ where $i = 0, 1, \ldots, \dim X$. Let b_i be a basis in $B_i = \mathrm{Im}\,(\partial \colon C_{i+1} \to C_i)$. Then

$$c_i' = \begin{cases} c_i, & i \neq k, k+1, \\ c_i, \hat{e}^k, & i = k, \\ c_i, \hat{e}^{k+1}, & i = k+1 \end{cases}$$

is a basis in $C_i' = C_i^\varphi(Y)$ and

$$b_i' = \begin{cases} b_i, & i \neq k, \\ b_i, \partial\hat{e}^{k+1}, & i = k \end{cases}$$

is a basis in $B_i' = \mathrm{Im}\,(\partial \colon C_{i+1}' \to C_i')$. (Note that $\partial\hat{e}^k = -\partial(\partial\hat{e}^{k+1} - \hat{e}^k) \in B_{k-1}$ so that $B_{k-1}' = B_{k-1}$.) Hence

$$\frac{\tau(C')}{\tau(C)} = \left(\frac{[b_{k+1}' b_k'/c_{k+1}']}{[b_{k+1} b_k/c_{k+1}]} \right)^{(-1)^k} \left(\frac{[b_k' b_{k-1}'/c_k']}{[b_k b_{k-1}/c_k]} \right)^{(-1)^{k+1}}$$

$$= \left(\frac{[b_{k+1} b_k \hat{e}^{k+1}/c_{k+1}\hat{e}^{k+1}]}{[b_{k+1} b_k/c_{k+1}]} \right)^{(-1)^k} \left(\frac{[b_k \partial\hat{e}^{k+1} b_{k-1}/c_k \hat{e}^k]}{[b_k b_{k-1}/c_k]} \right)^{(-1)^{k+1}}.$$

The first factor is clearly $+1$. Set $A = (b_k b_{k-1}/c_k)$. Since the incidence number of $\hat{e}^k$ and $\hat{e}^{k+1}$ is $+1$, the second factor equals

$$\left(\frac{(-1)^{\#b_{k-1}} \det \begin{pmatrix} A & 0 \\ * & 1 \end{pmatrix}}{\det A} \right)^{(-1)^{k+1}} = (-1)^{\#b_{k-1}} = (-1)^{\gamma_{k-1}(C)}$$

where the equality

$$\#b_{k-1} = \dim B_{k-1} = \gamma_{k-1}(C)$$

follows from the acyclicity of C and formula (2.1).

Next, set $E = C_*(X;\mathbb{R})$ and $E' = C_*(Y;\mathbb{R})$. The chain complexes E and E' come along with the bases e and $e' = e \cup \{e^k, e^{k+1}\}$. Choose a basis h of $H_*(E) = H_*(E')$ defining the given homology orientation. Set $\#b_i^0 = \dim \mathrm{Im}\,(\partial\colon E_{i+1} \to E_i)$. Proceeding as above, we find that

$$\frac{\mathrm{sign}\,(\check{\tau}(E',e',h))}{\mathrm{sign}\,(\check{\tau}(E,e,h))} = \frac{(-1)^{N(E')}}{(-1)^{N(E)}}(-1)^{\#h_k + \#b_{k-1}^0}.$$

The formula $\#e_i = \#h_i + \#b_i^0 + \#b_{i-1}^0$, where $i = 0,\ldots,\dim X$, yields $\gamma_{k-1}(E) = \beta_{k-1}(E) + \#b_{k-1}^0$. Hence,

$$(-1)^{\#h_k + \#b_{k-1}^0} = (-1)^{\gamma_{k-1}(E)+\beta_k(E)}.$$

Observe finally that $\beta_i(E') = \beta_i(E)$ for all i and

$$\gamma_i(E') = \begin{cases} \gamma_i(E), & i \neq k, \\ \gamma_i(E)+1, & i = k. \end{cases}$$

Hence,

$$N(E') = N(E) + \beta_k(E).$$

Putting everything together and using $\gamma_{k-1}(C) = \gamma_{k-1}(E)$ we find

$$\frac{\tau_\varphi^0(Y,\hat{e}',h)}{\tau_\varphi^0(X,\hat{e},h)} = (-1)^{N(E')-N(E)+\gamma_{k-1}(E)+\beta_k(E)+\gamma_{k-1}(C)}$$

$$= (-1)^{2(\beta_k(E)+\gamma_{k-1}(E))} = +1.$$

This finishes the proof of Theorem 18.3.   □

Theorems 8.8 and 18.3 imply that the sign-refined torsion $\tau_\varphi^0(X)$ is invariant under cellular subdivision of X. This allows us to apply this torsion to homology oriented compact pl-manifolds.

Applying the constructions of this section to the Milnor torsion $\tau_\mu(X)$ and to the maximal abelian torsion $\tau(X)$ of a homology oriented finite connected CW-complex X we obtain their sign-refined versions $\tau_\mu^0(X) \in Q(G)/G$ and $\tau^0(X) \in Q(H)/H$ where $H = H_1(X)$ and $G = H/\mathrm{Tors}\,H$. These torsions can be useful in the problem of finding simple homotopy equivalences $X \to X$ reversing the homology orientation. For instance, it is clear that if $\tau_\mu(X) \neq 0$, then any simple homotopy equivalence $X \to X$ which acts as the identity in $H_1(X)/\mathrm{Tors}\,H_1(X)$ preserves the homology orientation of X.

If M is a closed oriented manifold of odd dimension m, then M has a *canonical homology orientation* determined by any basis in $\oplus_{i<m/2}H_i(M;\mathbb{R})$ followed by the Poincaré dual basis in $\oplus_{i>m/2}H_i(M;\mathbb{R})$. The opposite orientation of M gives rise to the opposite homology orientation if and only if the number $\sum_{i<m/2}\dim H_i(M;\mathbb{R})$ is odd. We obtain the following corollary.

Corollary 18.5 *Let M be a closed connected oriented pl-manifold of odd dimension m with non-zero Alexander function and with odd $\sum_{i<m/2} \dim H_i(M; \mathbb{R})$. Then any pl-homeomorphism $M \to M$ acting as the identity in $H_1(M)/$ Tors $H_1(M)$ preserves the orientation of M.*

One can similarly use the sign-refined maximal abelian torsion to determine the lens spaces admitting orientation reversing self-homeomorphisms. For simplicity we restrict ourselves to the 3-dimensional case. Consider a lens space $L(p; q_1, q_2)$, where $p \geq 2$ and q_1, q_2 are relative prime to p. Define $r_1 \in \mathbb{Z}/p\mathbb{Z}$ by $r_1 q_1 = 1 \pmod{p}$. As we have seen in the proof of Corollary 10.3, there is a diffeomorphism $L(p; q_1, q_2) = L(p; 1, r_1 q_2)$. Therefore, we can assume from the very beginning that $q_1 = 1$.

Theorem 18.6 *The lens space $L(p; 1, q)$ admits an orientation reversing homeomorphism if and only if $q^2 = -1 \pmod{p}$.*

Proof. Suppose first $p = 2$. Then $L(p; 1, q) = L(2; 1, 1) = \mathbb{R}P^3$. We have $q^2 = -1 \pmod{p}$, and the homeomorphism

$$\mathbb{R}P^3 \to \mathbb{R}P^3, \quad [x_0 : x_1 : x_2 : x_3] \mapsto [-x_0 : x_1 : x_2 : x_3]$$

is orientation reversing. Suppose now $p \geq 3$. Fix an orientation of $L = L(p; 1, q)$. It gives rise to a homology orientation of L which allows us to consider the sign-refined maximal abelian torsion $\tau^0(L) \in \mathbb{Q}[H]/H$ where $H = H_1(L)$. Let $f: L \to L$ be an orientation reversing homeomorphism. Then f inverses the homology orientation in L. Therefore, $f_*(\tau^0(L)) = -\tau^0(L)$ where f_* is the map $\mathbb{Q}[H]/H \to \mathbb{Q}[H]/H$ induced by the isomorphism $f_*: H \to H$.

Let T be the distinguished generator of H and suppose that $f_*(T) = T^k$, where $k \in \mathbb{Z}/p\mathbb{Z}$ is prime to p. Define $r \in \mathbb{Z}/p\mathbb{Z}$ by $rq = 1 \pmod{p}$. The computation of $\tau(L)$ in Example 13.1 and the equality $f_*(\tau^0(L)) = -\tau^0(L)$ imply that

$$(T^k - 1)(T^{kr} - 1) = -T^d(T - 1)(T^r - 1) \in \mathbb{Q}[H] \qquad (18.1)$$

for some $d \in \mathbb{Z}/p\mathbb{Z}$. Evaluating this equality at p'th roots of unity we obtain that for each p'th root of unity $\zeta \in \mathbb{C}$,

$$(\zeta^k - 1)(\zeta^{kr} - 1) = -\zeta^d(\zeta - 1)(\zeta^r - 1).$$

We conclude as in the proof of Theorem 10.6 that

$$\{\pm k, \pm kr\} = \{\pm 1, \pm r\} \pmod{p}.$$

Case I. $k = \pm 1$.

If $k = 1$, then (18.1) implies $1 = -T^d$, a contradiction.

If $k = -1$, then (18.1) implies $1 = -T^{d+r+1}$, a contradiction.

Case II. $k = \pm r$ and $r \neq \pm 1$.

If $k = r$ and $kr = 1$, then $1 = -T^d$, a contradiction.

If $k = r$ and $kr = -1$, then $r^2 = -1$ and $q^2 = -1 \pmod{p}$.

If $k = -r$ and $kr = 1$, then $r^2 = -1$ and $q^2 = -1 \pmod{p}$.

If $k = -r$ and $kr = -1$, then $1 = -T^{d+r+1}$, a contradiction.

Assume now that $q^2 = -1 \pmod{p}$. Using (3), (2) and (1) of Section 10 we find diffeomorphisms

$$L(p; 1, q) \rightarrow L(p; -1, q) \rightarrow L(p; q^2, q) \rightarrow L(p; q, q^2).$$

Finally, the identity map $S^3 \rightarrow S^3$ induces a diffeomorphism $L(p; q, q^2) \rightarrow L(p; 1, q)$. It is easy to check that the composition $L(p; 1, q) \rightarrow L(p; 1, q)$ of these diffeomorphisms is orientation reversing. This completes the proof of Theorem 18.6.   □

19   The Conway link function

Three oriented links L_+, L_-, L_0 in S^3 constitute a *Conway triple* if they coincide outside a certain ball and look as in Figure 19.1 inside this ball.

L_+ L_- L_0

Figure 19.1: A Conway triple.

Examples of Conway triples will be given below.

Definition 19.1 A *one-variable Conway function* associates to any oriented link $L = L_1 \cup \cdots \cup L_l$ a rational function in one variable t, $\nabla_L(t) \in \mathbb{Q}(t)$, so that the following four axioms are satisfied:

(1) $\nabla_L(t)$ is invariant under ambient isotopies of L.

(2) If L is an unknot, then $\nabla_L(t) = (t - t^{-1})^{-1}$.

(3) If $l \geq 2$, then $\nabla_L(t) \in \mathbb{Z}[t^{\pm 1}]$.

(4) If L_+, L_-, L_0 is a Conway triple, then

$$\nabla_{L_+}(t) = \nabla_{L_-}(t) + (t - t^{-1})\nabla_{L_0}(t).$$

The identity in (4) is called the *Conway skein relation*.

Theorem 19.2 [7] *There exists a unique one-variable Conway function.*

Proof of uniqueness. Let δ be the difference between two one-variable Conway functions. We first show that δ_L vanishes for trivial links. For an unknot, this follows from (2). Suppose that $\delta_L = 0$ for all trivial links with $l-1$ components. Then (1) and (4) show that $\delta_L = 0$ for any trivial link with l components (see Figure 19.2 for the case $l = 2$).

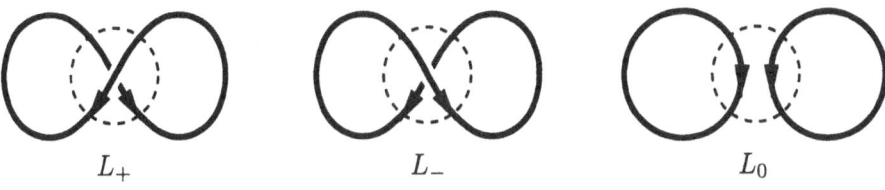

$$L_+ \qquad\qquad L_- \qquad\qquad L_0$$

Figure 19.2

Suppose now that $\delta_L = 0$ for links having diagrams with at most n crossings, and let L be a link given by a diagram with $n + 1$ crossings. By replacing successively certain undercrossings by overcrossings in this diagram, we obtain a diagram of a trivial link. By the skein relation and the inductive assumption, the value of δ is unchanged during these operations. Hence, $\delta_L = 0$.   □

For a simple proof of existence of the one-variable Conway function, we refer to [19]. The following exercises show how easily the one-variable Conway function can be computed.

Exercises 19.3 1. A link $L \subset S^3$ is called *splittable* if it can be separated into two non-void sublinks by a 2-sphere embedded in $S^3 \setminus L$. Show that $\nabla_L = 0$ for any splittable link L.

2. Let L be the Hopf link oriented as in Figure 19.3. Show that $\nabla_L = 1$.

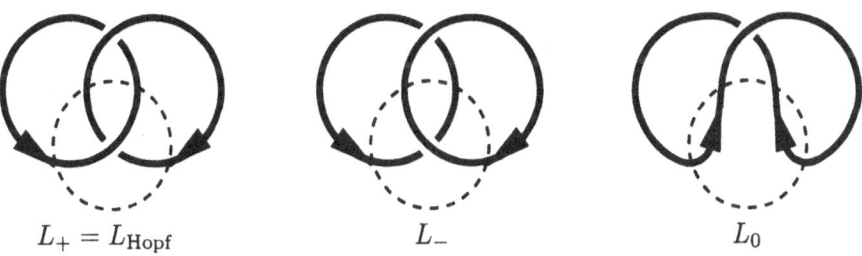

$$L_+ = L_{\text{Hopf}} \qquad\qquad L_- \qquad\qquad L_0$$

Figure 19.3

3. Let L be the Hopf link oriented as in Figure 19.3. Show that if the orientation of one of the components of L is reversed, then ∇ changes sign.

4. Use the Conway triples drawn in Figure 19.4 to prove that

$$\nabla_{L_{\text{trefoil}}} = \frac{t^2 - 1 + t^{-2}}{t - t^{-1}},$$

$$\nabla_{L_{\text{figure-eight}}} = -\frac{t^2 - 3 + t^{-2}}{t - t^{-1}},$$

$$\nabla_{L_{\text{Whitehead}}} = t^2 - 2 + t^{-2}.$$

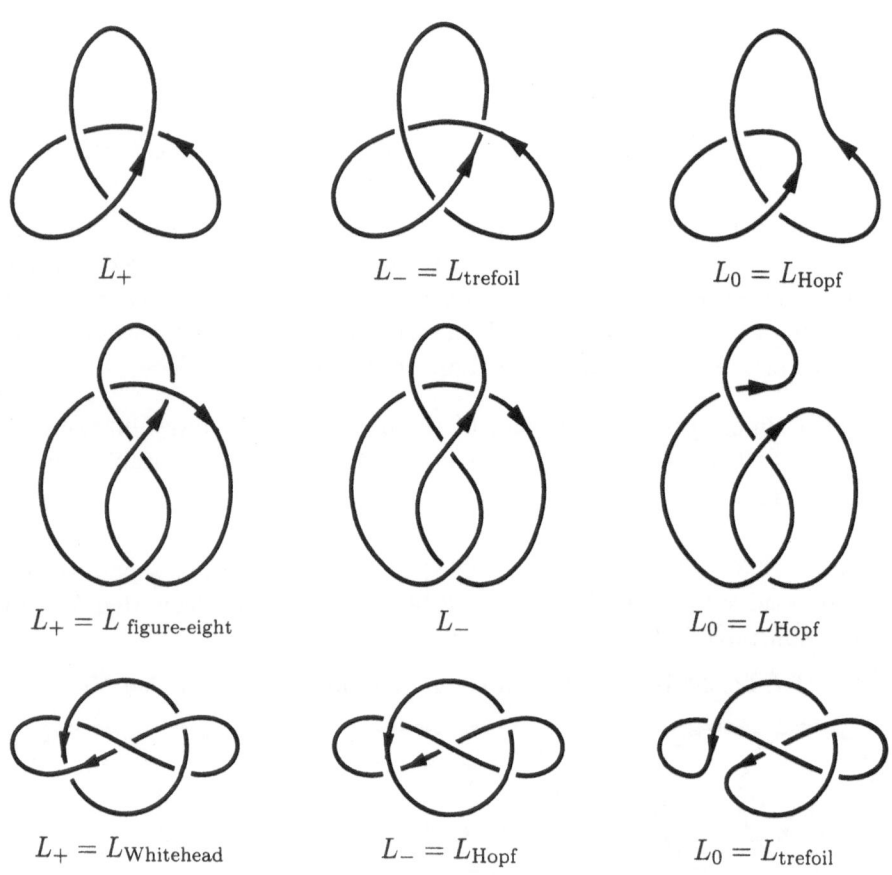

$$L_+ \qquad\qquad L_- = L_{\text{trefoil}} \qquad\qquad L_0 = L_{\text{Hopf}}$$

$$L_+ = L_{\text{ figure-eight}} \qquad\qquad L_- \qquad\qquad L_0 = L_{\text{Hopf}}$$

$$L_+ = L_{\text{Whitehead}} \qquad\qquad L_- = L_{\text{Hopf}} \qquad\qquad L_0 = L_{\text{trefoil}}$$

Figure 19.4

5. Show that for the Whitehead link L with any orientation

$$\nabla_L = t^2 - 2 + t^{-2} \qquad \text{and} \qquad \nabla_{L^*} = -(t^2 - 2 + t^{-2}),$$

where L^* is the mirror image of L. Deduce that the unoriented Whitehead link and its mirror are not ambiently isotopic.   ◇

In the following definition we need the notion of a (p, q)-cable of an oriented knot $K \subset S^3$, where p, q are mutually prime integers. Let U be a regular neighbourhood of K. A (p, q)-cable of K is an oriented knot $K' \subset \partial U \subset S^3$ homotopic to the p'th power of K in U and such that $\text{lk}(K, K') = q$. The knot K' is determined by K, p, q up to isotopy.

Definition 19.4 A *multivariable Conway function* associates to any ordered oriented link $L = L_1 \cup \cdots \cup L_l$ a rational function in l variables $t_1, \ldots, t_l$,

$$\widetilde{\nabla}_L(t_1, \ldots, t_l) \in \mathbb{Q}(t_1, \ldots, t_l),$$

so that the following four axioms are satisfied:

(1̃) $\widetilde{\nabla}_L(t_1, \ldots, t_l)$ is invariant under ambient isotopies of L.

(2̃) The function $\widetilde{\nabla}_L(t, \ldots, t)$ is the one-variable Conway function. In particular, it does not depend on the numbering of the components of L.

(3̃) If $l \geq 2$, then $\widetilde{\nabla}_L(t_1, \ldots, t_l) \in \mathbb{Z}[t_1^{\pm 1}, \ldots, t_l^{\pm 1}]$.

(4̃) If the link L' is obtained from L by replacing the i'th component L_i by its $(2, 1)$-cable, then

$$\widetilde{\nabla}_{L'}(t_1, \ldots, t_l) = (T + T^{-1})\, \widetilde{\nabla}_L(t_1, \ldots, t_{i-1}, t_i^2, t_{i+1}, \ldots, t_l),$$

where $T = t_i \prod_{j \neq i} t_j^{\text{lk}(L_i, L_j)}$.

Theorem 19.5 (Conway[7]) *There exists a unique multivariable Conway function.*

We outline a proof of Theorem 19.5 based on sign-refined torsions. We follow [34]; for another proof close to the original approach of Conway, see [17]. We begin with the uniqueness, which is the easy part of the theorem.

Proof of uniqueness. Let $\tilde{\delta}$ be the difference between two multivariable Conway functions. Clearly, $\tilde{\delta}$ satisfies (1̃), (3̃) and (4̃). By (2̃), the function $\delta_L(t) = \tilde{\delta}_L(t, \ldots, t)$ vanishes for each link L. If L is a knot, then $\tilde{\delta}_L = \delta_L = 0$. Let $L = L_1 \cup \cdots \cup L_l$ be a link with $l \geq 2$ components. Suppose that $\tilde{\delta}_L \neq 0$. Then, by (3̃), $\tilde{\delta}_L$ is a non-zero Laurent polynomial. Clearly, there are natural powers of two $a_i = 2^{b_i}$ with $b_i \geq 1$ for all $i = 1, \ldots, l$ such that $\tilde{\delta}_L(t^{a_1}, \ldots, t^{a_l}) \neq 0$. Let L' be the link obtained from L by successive b_i-fold replacement of L_i by its $(2, 1)$-cable for $i = 1, \ldots, l$. By (4̃), the Laurent polynomial $\tilde{\delta}_{L'}(t_1, \ldots, t_l)$ is a product of $\tilde{\delta}_L(t_1^{a_1}, \ldots, t_l^{a_l})$ and several Laurent polynomials of type $T + T^{-1}$ where T is a monomial. By what we have proved above, $\tilde{\delta}_{L'}(t, \ldots, t) = \delta_{L'}(t) = 0$. Therefore $\tilde{\delta}_L(t^{a_1}, \ldots, t^{a_l}) = 0$. This contradiction shows that $\tilde{\delta}_L = 0$. $\qquad\square$

We now discuss the existence of a multivariable Conway function. Consider an ordered oriented link $L = L_1 \cup \cdots \cup L_l$. Let $U = \coprod_{i=1}^l U_i$ be a regular neighbourhood of L. Set $E = S^3 \setminus \mathrm{Int}(U)$. Recall that the group $H_1(E)$ is a free abelian group on l generators $t_1, \ldots, t_l$ represented by the meridians of $L_1, \ldots, L_l$, respectively.

We shall define $\tilde{\nabla}_L$ via the sign-refined Milnor torsion. Endow E with a homology orientation as follows. As in the proof of Corollary 11.9 we see that E can be deformed onto a 2-dimensional subcomplex. Hence $H_3(E) = 0$. Since also $H_3(U) = H_2(U) = H_2(S^3) = 0$, the Mayer–Vietoris sequence of the triple (S^3, U, E) yields a short exact sequence

$$0 \to H_3(S^3) \to H_2(\partial U) \xrightarrow{\gamma} H_2(E) \to 0$$

The fixed orientation of S^3 induces an orientation of $U \subset S^3$ and hence an orientation of $\partial U = \coprod_{i=1}^l \partial U_i$. We have $H_3(S^3) = \mathbb{Z}$ and $H_2(\partial U) = \oplus_{i=1}^l \mathbb{Z}[\partial U_i]$. Since $\partial U = -\partial E$, the class $[\partial U] = [\partial U_1] + \cdots + [\partial U_l]$ is mapped by γ to $0 \in H_2(E)$. Hence, $[\partial U]$ generates $\mathrm{Ker}\,\gamma = \mathbb{Z}$. We conclude that $[\partial U_1], \ldots, [\partial U_{l-1}]$ is a basis of $H_2(E) = \oplus_{i=1}^{l-1}\mathbb{Z}$. Let $\mathfrak{m}_i$ be a meridian of L_i. We orient $H_*(E; \mathbb{R})$ by declaring

$$([pt], [\mathfrak{m}_1], \ldots, [\mathfrak{m}_l], [\partial U_1], \ldots, [\partial U_{l-1}])$$

to be a positive basis. Observe that the classes $[\mathfrak{m}_i]$ depend on the link orientation, and so the homology orientation of E depends on the link orientation.

Let $\tau^0 = \tau^0(t_1, \ldots, t_l) \in Q(H)$ be a representative of the sign-refined Milnor torsion $\tau_\mu^0(E) \in Q(H)/H$ where $H = H_1(E)$. Corollary 14.6 and Theorem 11.10 imply that

$$\tau^0(t_1^{-1}, \ldots, t_l^{-1}) = \pm t_1^{\nu_1} \cdots t_l^{\nu_l}\, \tau^0(t_1, \ldots, t_l)$$

with $\nu_1, \ldots, \nu_l \in \mathbb{Z}$. (The sign $\pm$ here may be computed to be $(-1)^l$ but we shall not need this.) Set

$$\tilde{\nabla}_L(t_1, \ldots, t_l) = -t_1^{\nu_1} \cdots t_l^{\nu_l}\, \tau^0(t_1^2, \ldots, t_l^2), \qquad (19.1)$$

It is easy to check that $\tilde{\nabla}_L$ does not depend on the choice of the representative τ^0 of $\tau_\mu^0(E)$.

We claim that $\tilde{\nabla}_L$ satisfies all the axioms of a multivariable Conway function. The fact that ambiently isotopic links have homeomorphic exteriors and the invariance of torsions under homeomorphisms imply that $\widetilde{\nabla}_L$ satisfies axiom ($\tilde{1}$). Axiom ($\tilde{3}$) follows from Corollary 15.3. Let us also check that for an oriented unknot K we have $\widetilde{\nabla}_K(t) = (t - t^{-1})^{-1}$. Let E be the exterior of K. The homology class t of the meridian of K is a generator of $H_1(E)$. By definition of the homology orientation of the link exterior, $([pt], t)$ is a positive basis of $H_*(E; \mathbb{R})$. The manifold E is a solid torus $S^1 \times D^2$, which is simple homotopy equivalent to its core $S^1 \subset E$. We orient $H_*(S^1; \mathbb{R})$ by declaring $([pt], t)$ to be a positive basis. Then, by Theorem 18.3, $\tau_\mu^0(E) = \tau_\mu^0(S^1)$. As in Lemma 6.2, consider a CW-decomposition of S^1 with one 0-cell and one 1-cell. For $C = C_*(S^1; \mathbb{R})$ we have $N(C) = -1$, and so, going through the definition of $\tau_\mu^0(S^1)$, we find that $\tau_\mu^0(S^1) = -(t-1)^{-1} \in \mathbb{Q}(t)/\{t^n\}_{n \in \mathbb{Z}}$. Since $-(t-1)^{-1} = -t(-(t-1)^{-1})$, we obtain $\widetilde{\nabla}_K(t) = -t(-(t^2-1)^{-1}) = (t-t^{-1})^{-1}$, as claimed. For the verification of the remaining axioms we refer to [34, pp. 156–163]. $\qquad\square$

Corollary 19.6 *Let $\widetilde{\nabla}$ be the multivariable Conway function and let $L = L_1 \cup \cdots \cup L_l$ be an ordered oriented link. Denote the mirror of L by L^*. Given a permutation σ of $\{1, \ldots, l\}$, set $L_\sigma = L_{\sigma(1)} \cup \cdots \cup L_{\sigma(l)}$. Then*

1. $\overline{\widetilde{\nabla}_L} = (-1)^l \, \widetilde{\nabla}_L.$
2. $\widetilde{\nabla}_{L^*} = (-1)^{l+1} \, \widetilde{\nabla}_L.$
3. $\widetilde{\nabla}_{L_\sigma}(t_1, \ldots, t_l) = \widetilde{\nabla}_L(t_{\sigma(1)}, \ldots, t_{\sigma(l)}).$

Proof. It is a simple matter to verify that the functions $L \mapsto (-1)^l \, \overline{\widetilde{\nabla}_L}$ and $L \mapsto (-1)^{l+1} \, \widetilde{\nabla}_{L^*}$ satisfy the axioms for a multivariable Conway function. This implies the first two claims. The third one is verified similarly. $\qquad\square$

Remark 19.7 By (19.1), the multivariable Conway function and the sign-refined Milnor torsion of an oriented link are equivalent invariants.

Corollary 19.8 *Let L be an ordered oriented l-component link in S^3. Then*

$$\Delta_L(t_1^2, \ldots, t_l^2) \doteq \begin{cases} \widetilde{\nabla}_L(t)(t - t^{-1}) & \text{if} \quad l = 1, \\ \widetilde{\nabla}_L(t_1, \ldots, t_l) & \text{if} \quad l \geq 2. \end{cases} \qquad (19.2)$$

Corollary 19.8 follows from Corollary 15.3 and (19.1). It shows that the Conway function determines the Alexander polynomial Δ_L. Conversely, Δ_L determines $\widetilde{\nabla}_L$ up to sign. The function $\widetilde{\nabla}_L$ cannot be completely restored from Δ_L. E.g., by Corollary 19.6.2, $\widetilde{\nabla}_{L^*} = (-1)^{l+1} \widetilde{\nabla}_L$ while $\Delta_{L^*} = \Delta_L$.

20 Euler structures

In this section we briefly present the theory of Euler structures on CW-complexes and smooth manifolds and their torsions, cf. [35].

20.1 Combinatorial Euler structures

Let X be a finite connected CW-complex. A family $\hat{e} = \{\hat{e}_i\}$ of open cells in the maximal abelian covering $\widehat{X}$ of X will be called *fundamental* if over each open cell e_i in X lies exactly one cell $\hat{e}_i$ of $\hat{e}$. Given two fundamental families $\hat{e}$ and $\hat{e}'$, we set

$$\hat{e}'/\hat{e} = \prod_{e_i \in X} (\hat{e}'_i/\hat{e}_i)^{(-1)^{\dim e_i}} \in H_1(X)$$

where $\hat{e}'_i/\hat{e}_i$ is the unique element $h \in H_1(X)$ such that $\hat{e}'_i = h\hat{e}_i$. Clearly,

$$\hat{e}/\hat{e} = 1, \quad \hat{e}/\hat{e}' = (\hat{e}'/\hat{e})^{-1}, \quad \hat{e}''/\hat{e} = (\hat{e}''/\hat{e}')(\hat{e}'/\hat{e}) \tag{20.1}$$

for any fundamental families $\hat{e}$, $\hat{e}'$, $\hat{e}''$. The relation

$$\hat{e} \sim \hat{e}' \quad \Longleftrightarrow \quad \hat{e}/\hat{e}' = 1 \tag{20.2}$$

is an equivalence relation on the set of fundamental families. The set of equivalence classes is denoted by $\mathrm{Eul}(X)$ and its elements are called *combinatorial Euler structures* on X.

If $h \in H_1(X)$ and e is the equivalence class of a fundamental family $\hat{e}$, let $h e$ be the class of any fundamental family $\hat{e}'$ with $\hat{e}'/\hat{e} = h$. It follows from (20.1) and (20.2) that this defines an action of $H_1(X)$ on $\mathrm{Eul}(X)$. It follows from definitions that this action is free and transitive. We conclude that the cardinality of $\mathrm{Eul}(X)$ equals the cardinality of $H_1(X)$.

Lemma 20.1 *For any cellular subdivision X' of X there is a canonical $H_1(X)$-equivariant bijection $\mathrm{Eul}(X') = \mathrm{Eul}(X)$.*

Proof. The maximal abelian covering $\widehat{X'}$ of X' is a cellular subdivision of $\widehat{X}$. If $\{\hat{e}_i\}$ is a fundamental family of open cells in $\widehat{X}$, then the open cells in $\widehat{X'}$

$$\left\{ \hat{f}_j \,\middle|\, \hat{f}_j \subset \hat{e}_i \text{ for some } i \right\}$$

form a fundamental family of cells in $\widehat{X'}$. It is easy to see that the assignment $\{\hat{e}_i\} \mapsto \{\hat{f}_j\}$ induces a map $\mathrm{Eul}(X) \to \mathrm{Eul}(X')$ which is $H_1(X)$-equivariant. Since the action of $H_1(X)$ on both $\mathrm{Eul}(X)$ and $\mathrm{Eul}(X')$ is free and transitive, this map is bijective. $\square$

Let $\mathbb{F}$ be a field and let $\varphi\colon \mathbb{Z}[H_1(X)] \to \mathbb{F}$ be a ring homomorphism such that $H_*^\varphi(X) = 0$. It follows from the computations in Section 6.1 that if $\hat{e}$ and $\hat{e}'$ are two fundamental families of cells in $\widehat{X}$, then

$$\tau_\varphi(X, \hat{e}') = \pm\varphi(\hat{e}'/\hat{e})\, \tau_\varphi(X, \hat{e}).$$

We may thus associate to each combinatorial Euler structure $\mathfrak{e}$ on X the torsion $\tau_\varphi(X, \mathfrak{e}) = \tau_\varphi(X, \hat{e}) \in \mathbb{F}/\{\pm 1\}$ where $\hat{e}$ is any fundamental family in the class $\mathfrak{e}$. If X is homology oriented then we can consider the sign-refined torsion $\tau_\varphi^0(X, \mathfrak{e}) \in \mathbb{F}$ which has no indeterminacy at all. By [35], these torsions are invariant under cellular subdivision of X.

We can apply the construction above to define the sign-refined maximal abelian torsion $\tau^0(X, \mathfrak{e}) \in Q(H_1(X))$ for any Euler structure $\mathfrak{e}$ on a homology oriented finite connected CW-complex X. It is defined by the formulas (13.1) and (13.2) in which we should now consider sign-refined torsions and choose the fundamental family $\hat{e}$ in the given class $\mathfrak{e}$. Note that for any $h \in H_1(X)$,

$$\tau^0(X, h\mathfrak{e}) = h\, \tau^0(X, \mathfrak{e}).$$

20.2 Geometric Euler structures

Consider a closed connected smooth m-dimensional manifold M with $\chi(M) = 0$. A continuous vector field on M is called *nonsingular* if it vanishes nowhere. According to the Poincaré–Hopf theorem, there exist nonsingular vector fields on M. We call two nonsingular vector fields u and v *equivalent*, $u \sim v$, if for some point $p \in M$ the restrictions of u and v to $M \setminus \{p\}$ are homotopic in the class of nonsingular vector fields on $M \setminus \{p\}$. Clearly, $\sim$ is an equivalence relation on the set of nonsingular vector fields on M. Denote the set of equivalence classes by $\mathrm{vect}(M)$. Elements of $\mathrm{vect}(M)$ are called *geometric Euler structures* on M. The element of $\mathrm{vect}(M)$ represented by a nonsingular vector field v on M is denoted by $[v]$.

If u, v are two non-singular vector fields on M, then the first obstruction to deforming u into v in the class of non-singular vector fields lies in $H^{m-1}(M) = H_1(M)$. This obstruction depends only on $[u], [v] \in \mathrm{vect}(M)$ and is denoted by $[u]/[v]$. It is easy to show that for any $[v] \in \mathrm{vect}(M)$, $h \in H_1(M)$ there is a unique Euler structure $h[v] \in \mathrm{vect}(M)$ such that $h[v]/[v] = h$. Thus, $H_1(M)$ acts freely and transitively on $\mathrm{vect}(M)$. We give here a geometric description of a nonsingular vector field in the class $h[v]$. Choose an oriented simple closed curve l in M which represents h. Let V be a regular neighbourhood of l. We endow V with coordinates (ϑ, r), where $\vartheta \in \mathbb{R}/2\pi\mathbb{Z}$ is the coordinate along l and r is the radial coordinate on the $(m-1)$-balls transversal to l. Choose a Riemannian metric on V such that the tangent vector fields $\frac{\partial}{\partial \vartheta}$ and $\frac{\partial}{\partial r}$ are orthogonal everywhere. Denote by d the function on V which assigns to a point its distance from l. We may assume that $d_{|\partial V} = \pi$, and after applying a homotopy to v we may assume that $v_{|V} = -\frac{\partial}{\partial \vartheta}$. Then $h[v]$ is represented by the nonsingular vector field on M which is equal to v on $M \setminus V$ and is equal to $\cos d \frac{\partial}{\partial \vartheta} + \sin d \frac{\partial}{\partial r}$ on V.

There is an involution on $\mathrm{vect}(M)$ sending the class $\mathfrak{e} = [v]$ of a non-singular vector field v to the class of the opposite vector field $-v$. Set $\mathfrak{e}^{-1} = [-v] \in \mathrm{vect}(M)$. One can check that $(h\mathfrak{e})^{-1} = h^{(-1)^m} \mathfrak{e}^{-1}$ for any $h \in H_1(M)$.

For every $\mathfrak{e} \in \mathrm{vect}(M)$ we define its *Euler class* $c(\mathfrak{e}) \in H_1(M)$ by $c(\mathfrak{e}) = \mathfrak{e}/\mathfrak{e}^{-1}$. If $\mathfrak{e} = [v]$ is the class of a non-singular vector field v, then $c(\mathfrak{e})$ can be computed to be the Poincaré dual of the Euler class of the $(m-1)$-dimensional vector bundle $v^{\perp}$ on M formed by the tangent vectors orthogonal to v. Note that $c(h\mathfrak{e}) = h^{1-(-1)^m} c(\mathfrak{e})$ for any $h \in H_1(M)$.

We set $\mathrm{Eul}(M) = \mathrm{Eul}(X)$ where X is any C^1-triangulation of M. By Lemma 20.1, this definition essentially does not depend on the choice of X.

Theorem 20.2 [35, 6.1] *Let M be a closed connected smooth manifold with $\chi(M) = 0$. Then there is a canonical $H_1(M)$-equivariant bijection $\mathrm{vect}(M) = \mathrm{Eul}(M)$.*

Sketch of a proof. Assume for concreteness that $m = \dim M \geq 3$. We start with recalling the *Whitney singular vector field* ν on M defined in terms of a C^1-triangulation X of M (cf. [16]). Let $A = \langle \underline{a}_0, \underline{a}_1, \dots, \underline{a}_k \rangle$ be a simplex of the first barycentric subdivision $X^{(1)}$ of X. Using the obvious linear structure on A we can write down any point $x \in A$ in the form

$$x = \sum_{i=0}^{k} \lambda_i(x) \, \underline{a}_i$$

where $\lambda_0(x), \dots, \lambda_k(x)$ are non-negative real numbers such that $\sum_{i=0}^{k} \lambda_i(x) = 1$. (These numbers are the *barycentric coordinates* of x.) Set

$$\nu(x) = \sum_{0 \leq i < j \leq k} \lambda_i(x) \, \lambda_j(x) \, (\underline{a}_i - x).$$

Here $\underline{a}_i - x$ is the tangent vector at $x \in A \subset M$ corresponding to the vector leading from x to $\underline{a}_i$ via the linear structure on A. The singular points of ν are precisely the barycenters of the simplices of X.

Now, let $p \colon \widehat{X} \to X$ be the maximal abelian covering. Let $\hat{e} = \{\hat{e}_i\}$ be a fundamental family of simplices in $\widehat{X}$. Fix a point $x_0 \in \widehat{X}$. For each i choose a path in $\widehat{X}$ leading from x_0 to the barycenter of $\hat{e}_i$. Projecting this path to X we obtain a path $\gamma_i \colon [0, 1] \to X$ leading from $p(x_0)$ to the barycenter of $e_i = p(\hat{e}_i)$. Since $m = \dim M \geq 3$, we can choose these paths so that they are embedded and disjoint except at their common endpoint $p(x_0)$. The union $\Gamma = \cup_i \gamma_i([0,1])$ is a wedge of intervals embedded in X. The 0-valent vertices of Γ are exactly the barycenters of the simplices in X. A regular neighbourhood U of Γ in M is an m-dimensional ball. The vector field ν is non-singular on $M \backslash U$. Since $\chi(M) = 0$, this non-singular vector field on $M \backslash U$ extends to a non-singular vector field on M. It can be shown that the class of this vector field in $\mathrm{vect}(M)$ depends only on the class of $\hat{e}$ in $\mathrm{Eul}(X) = \mathrm{Eul}(M)$. We thus obtain a map $\mathrm{Eul}(M) \to \mathrm{vect}(M)$. It is $H_1(M)$-equivariant and therefore bijective. $\qquad \square$

We use Theorem 20.2 to identify combinatorial and geometric Euler structures on manifolds. In particular, the torsions of combinatorial Euler structures defined in Section 20.1 can be applied to geometric Euler structures.

We now state a duality theorem for torsions of Euler structures following [34, 35]. For simplicity, we restrict ourselves to the case of closed oriented (and hence homology oriented) odd-dimensional manifolds.

Theorem 20.3 [34, 35] *Let M be a closed connected oriented smooth manifold of odd dimension m. Let $\mathbb{F}$ be a field with involution $f \mapsto \bar{f}$ and $\varphi \colon \mathbb{Z}[H_1(M)] \to \mathbb{F}$ be a ring homomorphism such that $H_*^\varphi(M) = 0$ and $\varphi(h^{-1}) = \overline{\varphi(h)}$ for all $h \in H_1(M)$. Then for any $\mathfrak{e} \in \mathrm{Eul}(M) = \mathrm{vect}(M)$,*

$$\overline{\tau_\varphi^0(M,\mathfrak{e})} = (-1)^z \, \tau_\varphi^0(M,\mathfrak{e}^{-1})$$

where

$$z = \begin{cases} 0 & \text{if} \quad m \equiv 3 \ (\mathrm{mod}\,4), \\ \sum_{i=0}^{(m-1)/2} \dim H_i(M;\mathbb{R}) & \text{if} \quad m \equiv 1 \ (\mathrm{mod}\,4). \end{cases}$$

Using the equality $\mathfrak{e}^{-1} = (c(\mathfrak{e}))^{-1}\mathfrak{e}$ we can rewrite this theorem as

$$\overline{\tau_\varphi^0(M,\mathfrak{e})} = (-1)^z \, \overline{\varphi(c(\mathfrak{e}))^{-1}} \, \tau_\varphi^0(M,\mathfrak{e}).$$

Theorem 20.3 implies the following duality theorem for the maximal abelian torsion $\tau^0(M,\mathfrak{e}) \in Q(H_1(M))$ of an Euler structure $\mathfrak{e}$ on a closed oriented (and hence homology oriented) odd-dimensional manifold M.

Corollary 20.4 *Let M be a closed connected oriented smooth manifold of odd dimension m. Then for any $\mathfrak{e} \in \mathrm{Eul}(M)$,*

$$\overline{\tau^0(M,\mathfrak{e})} = (-1)^z \, \tau^0(M,\mathfrak{e}^{-1}) = (-1)^z \, \overline{c(\mathfrak{e})}^{-1} \, \tau^0(M,\mathfrak{e})$$

where z is as in Theorem 20.3 and the overbar denotes the involution in $Q(H_1(M))$ sending $h \in H_1(M)$ to h^{-1}.

We refer to [35] for Euler structures on manifolds with boundary and for applications of torsions to the classification of Euler structures on manifolds; see also [4, 11, 12].

21 Torsion versus Seiberg–Witten invariants

In this last section we sketch the relationships between torsions and Seiberg–Witten invariants of 3-manifolds following [36, 37].

21.1 The torsion function T

We consider a closed connected oriented 3-manifold M with $b_1(M) \geq 1$. We provide M with the homology orientation induced by the orientation of M as in Section 18. Set $H = H_1(M)$. The two theorems stated in this subsection can be proved by the methods discussed previously in these notes. We refer to [36, Theorems 4.1 and 4.2.3] for details.

Theorem 21.1 *If $b_1(M) \geq 2$, then $\tau^0(M, \mathfrak{e}) \in \mathbb{Z}[H] \subset Q(H)$ for any $\mathfrak{e} \in$* Eul(M).

By this theorem, for any $\mathfrak{e} \in$ Eul(M) we have $\tau^0(M, \mathfrak{e}) = \sum_{h \in H} q^{\mathfrak{e}}(h)\, h$, where $q^{\mathfrak{e}}(h) \in \mathbb{Z}$. We define the *torsion function*

$$T \colon \mathrm{Eul}(M) \to \mathbb{Z} \quad \text{by } T(\mathfrak{e}) = q^{\mathfrak{e}}(1).$$

Note that knowing T we can recover the torsion $\tau^0(M, \mathfrak{e})$ for all $\mathfrak{e} \in$ Eul(M). Namely, $\tau^0(M, \mathfrak{e}) = \sum_{h \in H} T(h^{-1}\mathfrak{e})\, h$. This easily follows from definitions and the equality $\tau^0(M, h\mathfrak{e}) = h\, \tau^0(M, \mathfrak{e})$ for all $h \in H$ and $\mathfrak{e} \in$ Eul(M).

Suppose now that $b_1(M) = 1$. We fix an element $t \in H$ whose image in $G = H/\operatorname{Tors} H$ generates the infinite cyclic group G. By Lemma 17.1, $1 - t$ is invertible in $Q(H)$. For each $\mathfrak{e} \in$ Eul(M) there is a unique integer $K = K(\mathfrak{e}, t)$ such that $c(\mathfrak{e}) = \mathfrak{e}/\mathfrak{e}^{-1} \in t^K \operatorname{Tors} H$. Observe that K is even. Indeed, according to a classical theorem of Stiefel, M is parallelisable. Choose a parallelisation $u = (u_1, u_2, u_3)$ of M and set $\mathfrak{e}_0 = [u_1]$. Clearly, $-u_1$ is homotopic to u_1 and so $c(\mathfrak{e}_0) = 1$. Then

$$c(\mathfrak{e}) = c((\mathfrak{e}/\mathfrak{e}_0)\,\mathfrak{e}_0) = (\mathfrak{e}/\mathfrak{e}_0)^2 c(\mathfrak{e}_0) = (\mathfrak{e}/\mathfrak{e}_0)^2$$

and so $K(\mathfrak{e}, t) \in 2\mathbb{Z}$.

Theorem 21.2 *Let $b_1(M) = 1$. Set $\Sigma = \sum_{h \in \operatorname{Tors} H} h \in \mathbb{Z}[H]$. Then for any $\mathfrak{e} \in$ Eul(M) and any $t \in H$ as above*

$$\tau^0(M, \mathfrak{e}) + \frac{K(\mathfrak{e}, t) + 2}{2}(1 - t)^{-1}\Sigma - (1 - t)^{-2}\Sigma \in \mathbb{Z}[H].$$

Let us call an element $h \in H$ *negative* if $h \in t^k \operatorname{Tors} H$ with $k < 0$. Denote by Λ the Novikov ring of H consisting of series $\sum_{h \in H} q(h)\, h$ with $q(h) \in \mathbb{Z}$ such that $q(h) = 0$ for all but finitely many negative h. The multiplication in

Λ is induced by the group operation in H. Clearly, $\mathbb{Z}[H] \subset \Lambda$. Observe that $(1 - t)\left(\sum_{i \geq 0} t^i\right) = 1$ in Λ. Hence, $1 - t$ is invertible in Λ. By Theorem 21.2 we can therefore expand the torsion $\tau^0(M, \mathfrak{e})$ as an element $\sum_{h \in H} q^{\mathfrak{e}}(h) h$ of Λ. We define the *torsion function*

$$T_t \colon \mathrm{Eul}(M) \to \mathbb{Z} \quad \text{by} \quad T_t(\mathfrak{e}) = q^{\mathfrak{e}}(1).$$

As above, we can reconstruct the torsion $\tau^0(M, \mathfrak{e})$ from T_t using the equality $T_t(h\mathfrak{e}) = q^{\mathfrak{e}}(h^{-1})$ for all $h \in H$. Note that $T_{ht} = T_t$ for $h \in \mathrm{Tors}\, H$ and $T_{t^{-1}}(\mathfrak{e}) = T_t(\mathfrak{e}) + K(\mathfrak{e}, t)/2$ for all $\mathfrak{e} \in \mathrm{Eul}(M)$.

Remark 21.3 The statement of Theorem 21.2 corrects the corresponding statement in [36], which contained a sign miscalculation.

21.2 Spin^c-structures

Recall that $SO(3) = SU(2)/\{\pm 1\} = U(2)/U(1)$, where $U(1)$ lies in $U(2)$ as the diagonal subgroup. The projection $U(2) \to SO(3)$ is a principal circle bundle over $SO(3)$.

Consider a closed oriented 3-manifold M. Endow M with a Riemannian metric and consider the corresponding principal $SO(3)$-bundle of oriented orthonormal frames $f_M \colon Fr \to M$. A Spin^c-*structure on* M is a lift of f_M to a principal $U(2)$-bundle. More precisely, a Spin^c-structure on M is an isomorphism class of a pair (F, α), where $F \to M$ is a principal $U(2)$-bundle and α is an isomorphism of the principal $SO(3)$-bundle $F/U(1) \to M$ onto $f_M \colon Fr \to M$. The notion of a Spin^c-structure on M is essentially independent of the choice of a Riemannian metric on M. Denote the set of Spin^c-structures on M by Spin$^c(M)$. Since M is parallelisable, the set Spin$^c(M)$ is not empty.

The group $H^2(M) = H^2(M; \mathbb{Z})$ acts on Spin$^c(M)$ as follows. Let $h \in H^2(M)$ and let $F \to M$ be a Spin^c-structure. Recall that h determines a principal $U(1)$-bundle $E \to M$. The group $U(1)$ acts on $E \times F$ in the diagonal way and so we obtain a principal $U(2)$-bundle $(E \times F)/U(1) \to M$. It turns out that this action of $H^2(M)$ on Spin$^c(M)$ is free and transitive. Since $H_1(M) = H^2(M)$, we obtain a free and transitive action of $H_1(M)$ on Spin$^c(M)$.

Every nonsingular vector field v on M gives rise to a Spin^c-structure on M. Indeed, the tangent bundle TM splits as $v^\perp \oplus \mathbb{R}v$. Since both TM and $\mathbb{R}v$ are oriented, so is $v^\perp$. The structure group of TM is thereby reduced to $U(1)$. Consider now the principal $U(2)$-bundle on M associated to the principal $U(1)$-bundle TM and the inclusion $U(1) = U(1) \oplus (1) \subset U(2)$. It turns out that the Spin^c-structure thus obtained does not depend on the representative v of $[v] \in \mathrm{vect}(M)$. This yields a canonical $H_1(M)$-equivariant bijection $\mathrm{vect}(M) = \mathrm{Spin}^c(M)$ (cf. [36, Lemma 1.4]).

21.3 $SW = \pm T$

Consider a closed connected oriented 3-manifold M with $b_1(M) \geq 1$. Combining the bijections $\mathrm{Spin}^c(M) = \mathrm{vect}(M) = \mathrm{Eul}(M)$ with the torsion function $T \colon \mathrm{Eul}(M) \to \mathbb{Z}$ we obtain a map $\mathrm{Spin}^c(M) \to \mathbb{Z}$ denoted T_M. (In the case $b_1(M) = 1$ the maps T and T_M depend on the choice of an element $t \in H_1(M)$ whose image in $H_1(M)/\mathrm{Tors}\, H_1(M) \cong \mathbb{Z}$ is a generator.) There is another function $SW_M \colon \mathrm{Spin}^c(M) \to \mathbb{Z}$, which in the case $b_1(M) = 1$ also depends on $t \in H_1(M)$. The definition of SW runs parallel to the definition of the Seiberg–Witten invariant of 4-manifolds: One counts the gauge equivalence classes of solutions of the Seiberg–Witten equations, see, for instance, [22].

Theorem 21.4 [37] *For any closed connected oriented 3-manifold M with $b_1(M) \geq 1$ we have $SW_M = \pm T_M \colon \mathrm{Spin}^c(M) \to \mathbb{Z}$.*

Note that the sign $\pm$ may depend on M. The proof of Theorem 21.4 given in [37] is indirect. First, certain axioms for an abstract numerical invariant of Spin^c-structures on 3-manifolds are formulated. It is then shown that there exists at most one (up to sign) invariant satisfying these axioms. Finally, it is observed that both SW and T satisfy the axioms.

Theorem 21.4 can be used to study properties of the function $SW_M = \pm T_M$. For instance, consider the case $b_1(M) = 1$ and choose $e \in \mathrm{Eul}(M) = \mathrm{Spin}^c(M)$ such that $c(e) = 1$. It then follows from the definition of T_M, Theorem 21.2 and the equalities

$$(1-t)^{-1} = \sum_{i \geq 0} t^i, \qquad (1-t)^{-2} = \sum_{i \geq 0} (i+1)t^i$$

that for all $h \in t^k \mathrm{Tors}\, H_1(M)$ with sufficiently large $k > 0$, we have

$$T_M(he) = q^e(h^{-1}) = 0, \qquad T_M(h^{-1}e) = q^e(h) = -k.$$

The last formula shows that in this case SW_M has infinite support. In the case $b_1(M) \geq 2$, however, the function $SW_M \colon \mathrm{Spin}^c(M) \to \mathbb{Z}$ has finite support. This allows us to consider the sum of its values, denoted $|SW_M|$. By Theorem 21.4, $|SW_M|$ is equal (up to sign) to the sum of coefficients of $\tau(M) \in \mathbb{Z}[H_1(M)]/\pm H_1(M)$. By Lemma 13.2 and Theorem 14.12, this sum coincides with the sum of coefficients of the Alexander polynomial of M. This is nothing but the Casson-Walker-Lescop invariant of M (see [20]). In particular, if $b_1(M) \geq 4$, then $|SW_M| = 0$. If $b_1(M) = 3$, then

$$|SW_M| = \pm |\mathrm{Tors}\, H_1(M)|\, ((a_1 \cup a_2 \cup a_3)([M]))^2$$

where a_1, a_2, a_3 is an arbitrary basis of $H^1(M;\mathbb{Z}) = \mathbb{Z}^3$ and $\cup$ denotes the standard cup-product in cohomology. As an exercise, the reader may prove that for the 3-dimensional torus $M = S^1 \times S^1 \times S^1$ the function $SW_M = \pm T_M$ takes the value ± 1 on one element $e \in \text{vect}(M)$ and the value 0 on all other elements. The element e is represented by the "constant" vector field, i.e., the nonsingular vector field invariant under left and right translations on the Lie group $S^1 \times S^1 \times S^1$.

Theorem 21.4 gives a combinatorial computation of the Seiberg–Witten invariants of 3-manifolds (up to sign). Finding a combinatorial computation of the SW-invariants in dimension 4 is an outstanding challenge.

References

[1] J. Birman. Braids, links, and mapping class groups. Annals of Mathematics Studies, No. 82. Princeton University Press, Princeton, N.J. 1974.

[2] R. C. Blanchfield. Intersection theory of manifolds with operators with applications to knot theory. *Ann. of Math.* **65** (1957) 340–356.

[3] G. Burde, H. Zieschang. Knots. de Gruyter, Berlin, 1985.

[4] D. Burghelea. Removing metric anomalies from Ray–Singer torsion. *Lett. Math. Phys.* **47** (1999) 149–158.

[5] T. A. Chapman. Topological invariance of Whitehead torsion. *Amer. J. Math.* **96** (1974) 488–497.

[6] M. M. Cohen. A course in simple-homotopy theory. Graduate Texts in Mathematics, Vol. 10. Springer-Verlag, New York-Berlin, 1973.

[7] J. H. Conway. An enumeration of knots and links, and some of their algebraic properties. In: Computational Problems in Abstract Algebra. Pergamon Press, Oxford (1970) 329–358.

[8] R. H. Crowell, R. H. Fox. Introduction to knot theory. Ginn and Co., Boston, 1963.

[9] G. de Rham, S. Maumary and A. Kervaire. Torsion et type simple d'homotopie. *Lecture Notes in Mathematics.* **48** (1967).

[10] R. D. Edwards. On the topological invariance of simple homotopy type for polyhedra. *Amer. J. Math.* **100** (1978) 667–683.

[11] M. Farber, V. Turaev. Absolute torsion. *Contemp. Math.* **231** (1999) 73–85.

[12] M. Farber, V. Turaev. Poincaré–Reidemeister metric, Euler structures, and torsion. To appear in *J. Reine Angew. Math.*

[13] G. M. Fisher. On the Group of all Homeomorphisms of a Manifold. *Trans. Amer. Math. Soc.* **97** (1960) 193–212.

[14] W. Franz. Über die Torsion einer Überdeckung. *J. Reine Angew. Math.* **173** (1935) 245–254.

[15] W. Franz. Torsionsideale, Torsionsklassen und Torsion. *J. Reine Angew. Math.* **176** (1937) 113–124.

[16] S. Halperin, D. Toledo. Stiefel–Whitney homology classes. *Ann. of Math.* **96** (1972) 511–525.

[17] R. Hartley. The Conway potential function for links. *Comment. Math. Helv.* **58** (1983) 365–378.

[18] J. A. Hillman. Alexander ideals of links. *Lecture Notes in Mathematics.* **895** (1981).

[19] L. Kauffman. The Conway polynomial. *Topology* **20** (1981) 101–108.

[20] C. Lescop. Global surgery formula for the Casson–Walker invariant. Annals of Mathematics Studies, No. 140. Princeton University Press, Princeton, N.J. 1996.

[21] W. S. Massey. Algebraic Topology: An Introduction. Harcourt, New York, 1967.

[22] G. Meng and C. H. Taubes. $\underline{SW}$ = Milnor torsion. *Math. Res. Lett.* **3** (1996) 661–674.

[23] J. Milnor. A duality theorem for Reidemeister torsion. *Ann. of Math.* **76** (1962) 137–147.

[24] J. Milnor. Whitehead torsion. *Bull. Amer. Math. Soc.* **72** (1966) 358–426.

[25] J. Milnor. Infinite cyclic coverings. Topology of Manifolds (Michigan State Univ., E. Lansing, Mich., 1967) 115–133. Prindle, Weber & Schmidt, Boston, Mass..

[26] J. Munkres. Elementary differential topology. Annals of Mathematics Studies, No. 54. Princeton University Press, Princeton, N.J. 1963.

[27] D. B. Ray and I. M. Singer. R-Torsion and the Laplacian on Riemannian manifolds. *Adv. in Math.* **7** (1971) 145–210.

[28] K. Reidemeister. Homotopieringe und Linsenräume. *Hamburger Abhandl.* **11** (1935) 102–109.

[29] D. Rolfsen. Knots and links. Publish or Perish, Houston, 1990.

[30] C. P. Rourke and B. J. Sanderson. Introduction to piecewise-linear topology. Ergebnisse der Mathematik und ihrer Grenzgebiete, Band 69. Springer-Verlag, Berlin Heidelberg New York, 1972.

[31] H. Seifert. Über das Geschlecht von Knoten. *Math. Ann.* **110** (1934) 571–592.

[32] E. H. Spanier. Algebraic topology. Springer-Verlag, New York Heidelberg Berlin, 1966.

[33] V. G. Turaev. Reidemeister torsion and the Alexander polynomial. *Mat. Sb.* **101** (1976) 252–270.

[34] V. G. Turaev. Reidemeister torsion in knot theory. *Russian Math. Surveys* **41** (1986) 119–182.

[35] V. G. Turaev. Euler structures, nonsingular vector fields, and Reidemeister-type torsions. *Math. USSR-Izv.* **34** (1990) 627–662.

[36] V. G. Turaev. Torsion invariants of Spin^c-structures on 3-manifolds. *Math. Res. Lett.* **4** (1997) 679–695.

[37] V. G. Turaev. A combinatorial formulation for the Seiberg–Witten invariants of 3-manifolds. *Math. Res. Lett.* **5** (1998) 583–598.

[38] J. H. C. Whitehead. On C^1-complexes. *Ann. of Math.* **41** (1941) 809–824.

Index